Technische Universität München
Zentrum Mathematik

Uncertainty Principles on Riemannian Manifolds

Wolfgang Erb

Vollständiger Abdruck der von der Fakultät für Mathematik der Technischen Universität München zur Erlangung des akademischen Grades eines

Doktors der Naturwissenschaften (Dr. rer. nat.)

genehmigten Dissertation.

Vorsitzende:		Univ.-Prof. Dr. S. Rolles
Prüfer der Dissertation:	1.	Univ.-Prof. Dr. R. Lasser
	2.	Univ.-Prof. Dr. J. Prestin
		Universität zu Lübeck
	3.	Prof. Dr. J. Levesley
		University of Leicester
		(Schriftliche Beurteilung)

Die Dissertation wurde am 9.11.2009 bei der Technischen Universität München eingereicht und durch die Fakultät für Mathematik am 5.3.2010 angenommen.

Bibliografische Information der Deutschen Nationalbibliothek

Die Deutsche Nationalbibliothek verzeichnet diese Publikation in der
Deutschen Nationalbibliografie; detaillierte bibliografische Daten sind
im Internet über http://dnb.d-nb.de abrufbar.

ISBN 978-3-8325-2744-0

Logos Verlag Berlin GmbH
Comeniushof, Gubener Str. 47,
10243 Berlin
Tel.: +49 (0)30 42 85 10 90
Fax: +49 (0)30 42 85 10 92
INTERNET: http://www.logos-verlag.de

Abstract

In this thesis, the Heisenberg-Pauli-Weyl uncertainty principle on the real line and the Breitenberger uncertainty on the unit circle are generalized to Riemannian manifolds. The proof of these generalized uncertainty principles is based on an operator theoretic approach involving the commutator of two operators on a Hilbert space. As a momentum operator, a special differential-difference operator is constructed which plays the role of a generalized root of the radial part of the Laplace-Beltrami operator. Further, it is shown that the resulting uncertainty inequalities are sharp. In the final part of the thesis, these uncertainty principles are used to analyze the space-frequency behavior of polynomial kernels on compact symmetric spaces and to construct polynomials that are optimally localized in space with respect to the position variance of the uncertainty principle.

Zusammenfassung

In dieser Arbeit wird die Heisenberg-Pauli-Weyl'sche Unschärferelation und das Unschärfeprinzip von Breitenberger auf abstrakte Riemannsche Mannigfaltigkeiten verallgemeinert. Der Beweis dieses Unschärfeprinzips beruht auf einem operatortheoretischen Ansatz, in dem der Kommutator von zwei Operatoren auf einem Hilbertraum verwendet wird. Als Impulsoperator wird dabei ein spezieller Differential-Differenzenoperator konstruiert, der sich als verallgemeinerte Wurzel des radialen Teils des Laplace-Beltrami-Operators herausstellt. Ferner wird gezeigt, dass die resultierenden Ungleichungen scharf sind. Im letzten Teil der Arbeit werden die abgeleiteten Unschärfeprinzipien dazu benutzt um das Zeit-Frequenz-Verhalten von polynomiellen Kernen auf kompakten symmetrischen Räumen zu analysieren und Polynome zu konstruieren, die bezüglich der Ortsvarianz des Unschärfeprodukts optimal lokalisiert sind.

Non domandarci la formula che mondi possa aprirti,
sì qualche storta sillaba e secca come un ramo.
Codesto solo oggi possiamo dirti,
ciò che non siamo, ciò che non vogliamo.

Eugenio Montale, Non chiederci la
parola, Ossi di seppia, 1925

Contents

Introduction

In his famous work 'Über den anschaulichen Inhalt der quantentheoretischen Kinematik und Mechanik" (1927, [33]), Heisenberg revealed one of the fundamental principles of quantum mechanics, the uncertainty principle. This principle states that the values of two conjugate observables such as the position and the momentum of a quantum state f can not both be precisely determined at the same time. In particular, the more precisely one of the two properties is known, the less accurate the other variable can be measured. The most common way to describe the uncertainty principle mathematically is due to the following classical inequality, referred to as Heisenberg-Pauli-Weyl inequality (cf. [20]): If $f \in L^2(\mathbb{R})$ and $\|f\| = 1$, then

$$\int_{\mathbb{R}} (t - t_0)^2 |f(t)|^2 dt \cdot \int_{\mathbb{R}} |f'(t) - 2\pi i \omega_0 f(t)|^2 dt \geq \frac{1}{4}, \quad t_0, \omega_0 \in \mathbb{R}. \qquad (I.1)$$

In this formula, the quantum state f is interpreted as a L^2-density function on the real line $\mathbb{R}$. The value $\text{var}_S(f) := \int_{\mathbb{R}} (t - t_0)^2 |f(t)|^2 dt$ is the variance of the L^2-density f with respect to the mean value t_0 and is called the position variance of f. In view of the Fourier-Plancharel transform $\mathcal{F}(f)$ of the function f, the value $\text{var}_F(f) := \int_{\mathbb{R}} |f'(t) - 2\pi i \omega_0 f(t)|^2 dt$ corresponds to the position variance of $\mathcal{F}(f)$ in the frequency domain and is called the frequency or momentum variance of f. In this perspective, inequality (I.1) states that the product of the two variances $\text{var}_S(f)$ and $\text{var}_F(f)$ of the density f is always larger than $\frac{1}{4}$. Moreover, equality in (I.1) can be attained if and only if f corresponds to a Gaussian function, i.e., $f(t) = C e^{2\pi i \omega_0 t} e^{-\lambda(t - t_0)^2}$ where $C \in \mathbb{C}$ and $\lambda > 0$.

If the function f is defined on a manifold different from the real line $\mathbb{R}$, the question of how to formulate an uncertainty principle like (I.1) becomes more difficult. One main reason for this difficulty is due to the fact that for functions f on abstract manifolds it is not clear how appropriate position and frequency variances can be defined, nor it is clear whether Fourier techniques can be employed to determine a frequency variance.

To formulate an uncertainty principle on the unit circle $\mathbb{T}$, an interesting approach was pursued by Breitenberger in [6]. If one sets the frequency variance of a 2π-periodic function $f \in L^2([-\pi, \pi])$, $\|f\| = 1$, as

$$\text{var}_F(f) = \int_{-\pi}^{\pi} |f'(t)|^2 dt - \left| \int_{-\pi}^{\pi} f'(t) f(t) dt \right|^2 dt$$

1

and introduces a mean value $\varepsilon(f)$ by

$$\varepsilon(f) = \int_{-\pi}^{\pi} e^{it} |f(t)|^2 dt,$$

then it is possible to prove (cf. [6], [63], [68]) the following uncertainty principle:

$$(1 - |\varepsilon(f)|^2) \cdot \operatorname{var}_F(f) \geq \frac{1}{4} |\varepsilon(f)|^2. \tag{I.2}$$

As the Heisenberg-Pauli-Weyl inequality, also (I.2) has a physical interpretation. If one reads the value

$$\operatorname{var}_S(f) = \frac{1 - |\varepsilon(f)|^2}{|\varepsilon(f)|^2}$$

as the angular variance of the 2π-periodic density function f, then inequality (I.2) states that the values of the two observables angular position and angular momentum of f can not both be exactly determined at the same time. Furthermore, Prestin and Quak showed in [68] that the constant $\frac{1}{4}$ on the right hand side of inequality (I.2) is optimal.

Starting out from the classical inequalities (I.1) and (I.2), there have been a variety of attempts to generalize these uncertainty principles to more abstract settings. For the unit sphere $\mathbb{S}^d$, interesting results can be found in the papers of Rösler and Voit [73], Narcovich and Ward [62], Goh and Goodman [27] and Freeden and Windheuser [22]. Further, there exist several uncertainty principles for particular orthogonal expansions like the Jacobi polynomials [54, 73], the Bessel functions [74] as well as the Laguerre and Hermite polynomials [55]. Remarkable in this context is the fact that the classical inequalities (I.1) and (I.2) as well as the uncertainty inequalities in the above mentioned papers are proven by a related operator theoretic approach. Hereby, one defines two appropriate operators A and B in a Hilbert space $\mathcal{H}$. Then, the commutator $[A, B] = AB - BA$ is used to prove a simple Hilbert space inequality that provides the corner stone for the aimed-at uncertainty principle. For a brief summary of this approach, we refer to the survey articles of Folland and Sitaram [20] and Selig [79].

The aim of this thesis is to extend the uncertainty principles (I.1) and (I.2) to the more general setting of a Riemannian manifold M by means of the above mentioned operator theoretic approach. Similar as in (I.1) and (I.2), the obtained uncertainty principles contain an uncertainty product consisting of a position and a frequency variance term for which a general lower bound is derived. Of special interest for the proof of these uncertainty principles are the techniques developed by Rösler and Voit in [73]. In particular, in the course of Chapter 2, we will define a Dunkl operator that turns out to be a generalized root of the radial part of the Laplace-Beltrami operator Δ_M and that is used as a momentum operator to describe the frequency variance of a function $f \in L^2(M)$. If the manifold M is diffeomorphic to the Euclidean space, the resulting uncertainty principle (Theorem 2.54) will be similar to (I.1). On the other hand, if the manifold M is compact, we will use, as in (I.2), an appropriately introduced mean value $\varepsilon_p(f)$ to define a position variance with respect to a point $p \in M$ (see Corollary 2.43).

The proofs of the uncertainty principles are based on the following three ideas: First, one uses the exponential map and geodesic polar coordinates on the Riemannian manifold M to get an isometric isomorphism between the Hilbert space $L^2(M)$ and a weighted L^2-space on a cylindrical domain. Then, a symmetric extension of this weighted L^2-space is constructed and, thereon, an appropriate Dunkl operator is defined. Finally, this Dunkl operator together with a properly defined position operator and the commutator of these two operators is used to derive the uncertainty inequality. If M is diffeomorphic to the Euclidean space, then this proof leads to sharp uncertainty inequalities, where equality is attained for Gaussian-type functions on M. On the other side, if M is compact, then the obtained uncertainty inequality is asymptotically sharp (see Proposition 2.58).

Beside the uncertainty inequalities derived in this thesis, there exist a lot of uncertainty principles that are based on different approaches. In particular, we want to mention the articles [11, 72] of Ricci et al. and the article [56] of Martini in which uncertainty principles for general measure spaces have been worked out and that can be used also for Riemannian manifolds. As excellent summaries for a variety of well-known uncertainty principles, we refer also to the survey article [20] and the book [32] of Havin and Jöricke.

The main advantages of the utilized operator theoretic approach in this thesis are the sharpness of the resulting uncertainty principle and the availability of explicit expressions $\mathrm{var}_S(f)$ and $\mathrm{var}_F(f)$ for the position variance and the frequency variance of the function f. This turns the uncertainty principles developed in Chapter 1 and Chapter 2 into interesting auxiliary tools for space-frequency analysis. The variances $\mathrm{var}_S(f)$ and $\mathrm{var}_F(f)$ provide a good measure on how well a function f is localized in the space and the frequency domain and give substantial information on the space-frequency properties of possible wavelets and frames. For the classical case $M = \mathbb{R}^d$, there exists a broad theory on this subject and we refer to the monograph [29] as a fine introduction. Also on the unit circle, the Breitenberger uncertainty principle (I.2) provides a remarkable tool to study the space-frequency localization of trigonometric wavelets (cf. [63], [69] and [78]) or to construct optimally space localized trigonometric polynomials (see the article [71] of Rauhut). Similarly, a related formula for the position variance on the unit sphere $\mathbb{S}^d$ can be used to determine space optimal spherical harmonics (see the work [48] of Laín Fernández).

The objective of the last chapter in this thesis is to make use of the above mentioned advantages and to utilize the uncertainty principles developed in Chapter 1 and 2 for space-frequency analysis. We will present a few scenarios in which these uncertainty principles can be used to analyze the space-frequency behavior of particular polynomial kernels and to construct polynomials that are optimally localized in space with respect to the position variance. In particular, the theory of optimally space localized polynomials on the unit circle and on the unit sphere is extended to the more general setting of Jacobi expansions and compact two-point homogeneous spaces. These results are also related to the works of Filbir, Mhaskar and Prestin [17] and Petrushev and Xu [66] in which exponentially and sub-exponentially localized polynomials are constructed. Further, we will discuss the space-frequency behavior of the Christoffel-Darboux kernel which plays an important role in the theory of polynomial approximation, in particular on compact two-point homogeneous spaces (see the article [53] of Levesley and Ragozin). Finally, we will consider the de La Vallée Poussin kernel as an example of a polynomial kernel for which the uncertainty product tends to the optimal lower bound.

Outline of the thesis

In Chapter 1, we start out by giving a short overview on the existing theory of uncertainty principles in general Hilbert spaces (Theorems 1.2 and 1.4). As first examples of these principles based on an operator theoretic approach, we will encounter the Heisenberg-Pauli-Weyl uncertainty on the real line (Theorem 1.5) and the Breitenberger principle for 2π-periodic functions (Theorem 1.7). Then, based on an approach developed by Rösler and Voit [73] involving a particular Dunkl operator, uncertainty inequalities for functions in a weighted L^2-space are constructed. The considered underlying sets are the unit interval $[0, \pi]$ (Section 1.4.1) and the positive real half-axis (Section 1.4.3). Further, in Section 1.4.2, we consider an interesting new intermediate result where the functions in the uncertainty inequality have to satisfy a zero right-hand side boundary condition. Finally, in Section 1.5, we will present uncertainty principles for weighted L^2-spaces where the weight function is connected to a particular orthogonal expansion. Thereby, we will focus on functions that have an expansion in terms of Jacobi and Laguerre polynomials.

Chapter 2 contains the main new results of this thesis. In a first step, we will generalize the theory of Chapter 1 to a multi-dimensional setting. In particular, we will proof uncertainty principles for weighted L^2-spaces where the underlying set is a cylinder Z_π^d or a tube Z_∞^d of dimension d (Section 2.1). Using the exponential map on the tangent space T_pM of a Riemannian manifold M, we will construct an isometric isomorphism that maps the Hilbert space $L^2(M)$ onto such a weighted L^2-space. In this way, we are able to proof uncertainty principles for general Riemannian manifolds by using the theory of Section 2.1. We will distinguish between three different types of settings:

(1) In the first setting, the Riemannian manifold M is supposed to be compact. The obtained uncertainty principle in Theorem 2.41 can be considered as a generalization of the Breitenberger principle. As examples, we will encounter the d-dimensional spheres (Section 2.6.1), the projective spaces (Section 2.6.2) and the flat tori (Section 2.6.3).

(2) In the second case, we get an uncertainty inequality (Theorem 2.51) for functions f defined on a compact star-shaped subdomain Ω of a Riemannian manifold M with Lipschitz continuous boundary $\partial\Omega$ and the additional assumption that f vanishes at $\partial\Omega$. This uncertainty principle is a generalization of Theorem 1.24 in Section 1.4.2.

(3) In the third case, we will develop uncertainty principles for Riemannian manifolds that are diffeomorphic to the Euclidean space $\mathbb{R}^d$ (Theorem 2.54). It will turn out that these uncertainty principles are multi-dimensional analogs of the original Heisenberg-

Pauli-Weyl inequality. As an example, we will treat the hyperbolic space (Section 2.6.4).

In the third case, an important statement of Theorem 2.54 is the sharpness of the resulting uncertainty inequality. The proof that the uncertainty principle for compact Riemannian manifolds and the uncertainty principle for compact star-shaped domains are also asymptotically sharp, can be found in Section 2.5. Finally, in Section 2.7, we will investigate how information on the curvature of the Riemannian manifold M can be used to simplify the derived uncertainty inequalities. In particular, depending on the curvature, we will find easier to handle lower estimates of the original uncertainty inequalities. The main part of Chapter 2 is already published and can be found in the article [15].

In Chapter 3, we will use the uncertainty principle for Jacobi expansions on $[0, \pi]$ and the related uncertainty principle for compact two-point homogeneous spaces, i.e., the uncertainty principle for the spheres and the projective spaces developed in Section 2.6.1 and 2.6.2, to construct optimally space localized polynomials. In particular, we will determine those elements of a finite-dimensional polynomial subspace that minimize the position variance of the respective uncertainty principle.

In Section 3.1, we will develop a simple theory to determine the optimally space localized polynomials for Jacobi expansions on $[0, \pi]$ (Theorem 3.6). Using a generalized Christoffel-Darboux formula, we can state these optimal polynomials explicitly (Corollary 3.10). Further, we will compare the space-frequency localization of the optimally space localized polynomials with the space-frequency behavior of other well known polynomial kernels like the Christoffel-Darboux kernel and the de La Vallée Poussin kernel. It turns out that the uncertainty product of the Christoffel-Darboux kernel tends linearly to infinity as the order n of the polynomial kernel tends to infinity (Theorem 3.15), whereas the uncertainty product of the de La Vallée Poussin kernel tends to the optimal lower bound of the uncertainty principle (Theorem 3.16).

Section 3.2 contains some intermediate results to describe the behavior of the extremal zeros of associated Jacobi polynomials if one of the involved parameters is changing. In the final Section 3.3, these intermediate results are used to determine optimally space localized polynomials on compact two-point homogeneous spaces.

In Chapter A of the appendix, we give a short introduction into the general theory of Riemannian manifolds including the concepts of geodesics, the exponential map, curvature and integration on manifolds. In the second appendix Chapter B, there is a recapitulation of some basic facts concerning function spaces, the Stone-Weierstrass Theorem and operators in Hilbert spaces.

Acknowledgements

First of all, I sincerely thank to my advisors Prof. Rupert Lasser and Frank Filbir for their excellent support and backing during my studies in Munich. In his lectures on functional analysis and operator theory, Prof. Lasser taught me most of the basics I needed for this thesis. Many discussions with Frank on time-frequency analysis and polynomial approximation had a strong influence on main parts of this thesis. Moreover, the collaboration with Frank led to the joint article [16].

Many thanks go also to Ferenc Toókos with whom I developed the results of Section 3.2 and to Josef Obermaier who helped me in a lot of questions concerning orthogonal polynomials.

Further, I am very grateful to all colleagues and friends at the Institute of Biomathematics and Biometry at the Helmholtz Center Munich and at the Centre for Mathematical Sciences at the Munich University of Technology (TUM). In particular, I would like to thank all members of the Graduate Program Applied Algorithmic Mathematics.

Finally, my special thanks go to Thomas März, Eva Perreiter and Andreas Weinmann.

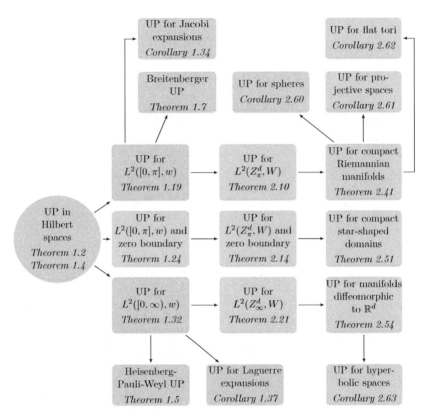

Figure 1: The main Theorems of Chapter 1 and 2. The parts highlighted in red are new in this thesis. UP = uncertainty principle.

*'The uncertainty principle "protects" quantum mechanics. Heisen-
berg recognized that if it were possible to measure the momentum
and the position simultaneously with a greater accuracy, the quantum mechanics would collapse. So he proposed that it must be
impossible. Then people sat down and tried to figure out ways
of doing it, and nobody could figure out a way to measure the
position and the momentum of anything - a screen, an electron, a
billiard ball, anything - with any greater accuracy.'*

R.P. Feynman, R.B. Leighton and M. Sands, The Feynman
Lectures on Physics III: Quantum Mechanics, Addison Wesley
Publishing Company, 1965, p. 1-11

1

Uncertainty principles - An overview

1.1. Uncertainty inequalities in Hilbert spaces

In this first section, we will present a very common approach to express uncertainty principles mathematically. It is based on a Hilbert space inequality involving the commutator
of two self-adjoint, or more generally, normal operators defined on the Hilbert space. The
various details of this theory and further references can be found in [19], [20], [28] and
[79]. A short introduction into the terminology of operators on Hilbert spaces can be
found in Section B.3 of the appendix.

A main advantage of the operator theoretic approach to uncertainty is the fact that
the uncertainty can be interpreted in terms of quantum mechanical observables. Let $\mathcal{H}$
denote a Hilbert space with inner product $\langle \cdot, \cdot \rangle$ and norm $\| \cdot \| = \sqrt{\langle \cdot, \cdot \rangle}$. Then, the
state of a quantum mechanical system is represented by a unit vector $v \in \mathcal{H}$, and the
observable quantity is usually represented by a self-adjoint operator A on $\mathcal{H}$. By the
spectral theorem, there exists an operator-valued measure P such that A decomposes
into $A = \int \lambda dP(\lambda)$. The map $\mu_v(E) = \langle P(E)v, v \rangle$ is a probability measure on $\mathbb{R}$ that
represents the distribution of the observable A in the state v. The mean and variance of
this measure are given by

$$\varepsilon_A(v) := \langle Av, v \rangle = \int_{\mathbb{R}} \lambda \langle dP(\lambda)v, v \rangle, \tag{1.1}$$

$$\mathrm{var}_A(v) := \| \left(A - \varepsilon_A(v) \right) v \|^2 = \| Av \|^2 - |\varepsilon_A(v)|^2 \tag{1.2}$$
$$= \int_{\mathbb{R}} (\lambda - \varepsilon_A(v))^2 \langle dP(\lambda)v, v \rangle.$$

The value $\varepsilon_A(v)$ is called the expectation value of the observable A and $\mathrm{var}_A(v)$ the variance of A in the state v. In general, the variance $\mathrm{var}_A(v)$ can be interpreted as a measure of the uncertainty of A in the state v and an uncertainty principle is an assertion about the product of two variances of two different observables A and B.

In the following, we suppose that the linear operators A and B are densely defined on $\mathcal{H}$, with domains $\mathcal{D}(A) \subset \mathcal{H}$ and $\mathcal{D}(B) \subset \mathcal{H}$. The domain for the product AB is given by

$$\mathcal{D}(AB) := \{v \in \mathcal{D}(B) : \ Bv \in \mathcal{D}(A)\}$$

and $\mathcal{D}(BA)$ likewise. If the commutator of the operators A and B is defined by

$$[A, B] := AB - BA \quad \text{on} \quad \mathcal{D}([A, B]) := \mathcal{D}(AB) \cap \mathcal{D}(BA),$$

the following uncertainty principle holds (cf. [19, Theorem 1.34]):

Theorem 1.1.
If A and B are self-adjoint operators on a Hilbert space $\mathcal{H}$, then

$$\|(A - a)v\| \cdot \|(B - b)v\| \geq \frac{1}{2}|\langle [A, B]v, v \rangle| \tag{1.3}$$

for all unit vectors $v \in \mathcal{D}([A, B])$ and $a, b \in \mathbb{R}$. Equality holds if and only if $(A - a)v$ and $(B - b)v$ are purely imaginary scalar multiplies of one another.

It is not always the case that the considered operators A and B are self-adjoint, for instance, when it is not possible to find a self-adjoint extension of a symmetric operator. Analyzing the proof of Theorem 1.34 in [19], one can see that the symmetry of the operators A and B suffices to prove the uncertainty principle. So, we get the following generalization (cf. [79, Theorem 3.3]):

Theorem 1.2.
If A and B are symmetric operators on $\mathcal{H}$, then

$$\|(A - a)v\| \cdot \|(B - b)v\| \geq \frac{1}{2}|\langle [A, B]v, v \rangle| \tag{1.4}$$

for all unit vectors $v \in \mathcal{D}([A, B])$ and $a, b \in \mathbb{R}$. Equality holds if and only if $(A - a)v$ and $(B - b)v$ are purely imaginary scalar multiplies of one another.

Proof. For $a, b \in \mathbb{R}$ and $v \in \mathcal{D}([A, B])$, we have

$$[(A - a), (B - b)]v = ABv - BAv = [A, B]v.$$

Now, the symmetry of the operators A and B implies

$$\begin{aligned}
\langle [A, B]v, v \rangle &= \langle (A - a)(B - b)v - (B - b)(A - a)v, v \rangle \\
&= \langle (B - b)v, (A - a)v \rangle - \langle (A - a)v, (B - b)v \rangle \\
&= 2i\,\mathrm{Im}(\langle (B - b)v, (A - a)v \rangle).
\end{aligned}$$

The imaginary part of $\langle [A, B]v, v \rangle$ is bounded from above by the absolute value of $\langle [A, B]v, v \rangle$. Applying the Schwarz inequality yields

$$\langle [A, B]v, v \rangle \leq 2|\langle (B - b)v, (A - a)v \rangle| \leq 2\|(A - a)v\| \cdot \|(B - b)v\|.$$

Equality holds for the Schwarz inequality if and only if $(A - a)v = \lambda(B - b)v$ for $\lambda \in \mathbb{C}$ and for the first inequality if and only if $\text{Re}\,\lambda = 0$. $\qquad\square$

Remark 1.3. The proof of inequality (1.4) is quite simple, but there are some subtleties hidden in the statement of Theorem 1.2. In fact, there are examples where the domain $\mathcal{D}([A, B])$ of the commutator gets very small or consists only of the zero vector. Moreover, the commutator $[A, B]$ is in general not closed and one can show that inequality (1.4) does usually not hold for vectors $v \in \mathcal{D}(\overline{[A, B]})$, where $\overline{[A, B]}$ denotes the closure of the operator $[A, B]$. For the detailed investigation of these counterexamples, we refer to [20]. Altogether, we can conclude that, when using this kind of uncertainty inequality, one has to keep a close watch at the domain $\mathcal{D}([A, B])$ of the commutator.

Now, one may ask for which choices of a and b the left hand sides of inequalities (1.3) and (1.4) are minimized. The answer is given by the Hilbert Projection Theorem. Namely, for $Av \in \mathcal{H}$ the point $v_0 \in \text{span}\{v\}$ with minimal distance to Av is exactly the orthogonal projection of Av on the one-dimensional linear subspace spanned by v, i.e. $v_0 = \langle Av, v \rangle v$. Therefore, if we take the variances

$$\text{var}_A(v) = \|(A - \varepsilon_A(v))v\|^2 = \min_{a \in \mathbb{R}} \|(A - a)v\|^2, \tag{1.5}$$

$$\text{var}_B(v) = \|(B - \varepsilon_B(v))v\|^2 = \min_{b \in \mathbb{R}} \|(B - b)v\|^2, \tag{1.6}$$

the inequalities (1.3) and (1.4) can be reformulated as

$$\text{var}_A(v) \cdot \text{var}_B(v) \geq \frac{1}{4}|\langle [A, B]v, v \rangle|^2. \tag{1.7}$$

Another interesting situation emerges when one of the two operators A or B is normal on $\mathcal{H}$, but not necessarily symmetric or self-adjoint. Also in this case, it is possible to formulate an uncertainty principle like (1.4) (cf. [79, Theorem 5.1]).

Theorem 1.4.
If B is symmetric and A is a normal operator on the Hilbert space $\mathcal{H}$, then

$$\|(A - a)v\| \cdot \|(B - b)v\| \geq \frac{1}{2}|\langle [A, B]v, v \rangle| \tag{1.8}$$

for all unit vectors $v \in \mathcal{D}([A, B])$ and $a \in \mathbb{C}$, $b \in \mathbb{R}$. Equality holds if and only if

$$(B - b)v = \lambda(A - a)v = -\bar{\lambda}(A^* - \bar{a})v$$

for a complex scalar $\lambda \in \mathbb{C}$.

Proof. If A is a normal operator, then also $A - aI$ is normal and $A^* - \bar{a}I$ is its adjoint operator. Moreover, $\mathcal{D}(A^*) = \mathcal{D}(A)$ and $\|(A - a)v\| = \|(A^* - \bar{a})v\|$ for all $v \in \mathcal{D}(A)$. Using the triangle and the Schwarz inequality, we get the estimate

$$|\langle [A, B]v, v\rangle| = |\langle (A - a)(B - b)v - (B - b)(A - a)v, v\rangle|$$
$$\overset{(*)}{\leq} |\langle (B - b)v, (A^* - \bar{a})v\rangle| + |\langle (A - a)v, (B - b)v\rangle|$$
$$\overset{(**)}{\leq} \|(B - b)v\| \cdot \|(A^* - \bar{a})v\| + \|(A - a)v\| \cdot \|(B - b)v\|$$
$$= 2\|(B - b)v\| \cdot \|(A - a)v\|.$$

For the Schwarz inequality in $(**)$, equality holds if and only if

$$(B - b)v = \lambda_1(A - a)v, \quad (B - b)v = \lambda_2(A^* - \bar{a})v, \quad \lambda_1, \lambda_2 \in \mathbb{C}.$$

The normality of A implies that $|\lambda_1| = |\lambda_2|$. Finally, the triangle inequality in $(*)$ becomes an equality if and only if $\lambda_1 = -\bar{\lambda}_2$. $\qquad\square$

1.2. The Heisenberg-Pauli-Weyl uncertainty principle

The first and undoubtedly most famous uncertainty principle goes back to Heisenberg and his pathbreaking work [33] of 1927. The mathematical version of this principle was formulated afterwards by Kennard [45] and by Pauli and Weyl (see [86], p. 77). It is the main example of an uncertainty inequality in a Hilbert space and can be found nowadays in numerous monographs and articles. As references, one may consider [19], [20], [29] and [79].

The underlying Hilbert space for the Heisenberg-Pauli-Weyl-principle is the space $L^2(\mathbb{R})$ of square integrable functions on the real axis $\mathbb{R}$ with inner product $\langle f, g\rangle := \int_{\mathbb{R}} f(t)\overline{g(t)}dt$ and norm $\|f\|^2 := \langle f, f\rangle$. As a position operator A and as a momentum operator B, we define

$$Af(t) := tf(t), \quad \mathcal{D}(A) := \{f \in L^2(\mathbb{R}) : tf \in L^2(\mathbb{R})\},$$
$$Bf(t) := if'(t), \quad \mathcal{D}(B) := \{f \in AC_{loc}(\mathbb{R}) : f' \in L^2(\mathbb{R})\}.$$

The set $AC_{loc}(\mathbb{R})$ denotes the space of all locally absolutely continuous functions on $\mathbb{R}$. For a brief summary on absolutely continuous functions, we refer to Section B.1 of the appendix. The operators A and B are densely defined in $L^2(\mathbb{R})$. Further, since

$$\int_{\mathbb{R}} tf(t)\overline{g(t)}dt = \int_{\mathbb{R}} f(t)\overline{tg(t)}dt \quad f, g \in \mathcal{D}(A),$$
$$\int_{\mathbb{R}} if'(t)\overline{g(t)}dt = -\int_{\mathbb{R}} f(t)\overline{ig'(t)}dt \quad f, g \in \mathcal{D}(B),$$

the operators A and B are both symmetric on their respective domains. Moreover, with a little bit of extra effort one can show that A and B are even self-adjoint (cf. [70, Lemma 2.2.1] and [85, Example (b), p. 318]). In terms of the Fourier-Plancharel transform

$$\mathcal{F}(f)(\omega) := \int_{\mathbb{R}} f(t)e^{-2\pi i \omega t}dt, \quad \mathcal{F}^{-1}(g)(t) := \int_{\mathbb{R}} g(\omega)e^{2\pi i \omega t}d\omega,$$

the momentum operator B can be reformulated as

$$Bf(t) = (\mathcal{F}^{-1}A\mathcal{F}(f))(t).$$

In this way, we get the following version of the Heisenberg-Pauli-Weyl principle:

Theorem 1.5 (Heisenberg-Pauli-Weyl uncertainty principle).
Let $f \in AC_{loc}(\mathbb{R}) \cap L^2(\mathbb{R})$ with $tf, f', tf' \in L^2(\mathbb{R})$, $\|f\| = 1$ and

$$t_0 := \int_{\mathbb{R}} t|f(t)|^2 dt, \quad \omega_0 := \int_{\mathbb{R}} \omega|\mathcal{F}(f)(\omega)|^2 d\omega.$$

Then,

$$\left(\|tf\|^2 - |\langle tf, f \rangle|^2\right) \cdot \left(\|f'\|^2 - |\langle f', f \rangle|^2\right) \geq \frac{1}{4}, \tag{1.9}$$

or equivalently

$$\int_{\mathbb{R}} (t - t_0)^2 |f(t)|^2 dt \cdot \int_{\mathbb{R}} (\omega - \omega_0)^2 |\mathcal{F}(f)(\omega)|^2 d\omega \geq \frac{1}{16\pi^2}. \tag{1.10}$$

Equality is attained if and only if $f(t) = Ce^{2\pi i \omega_0 t}e^{-\lambda(t-t_0)^2}$ for $C \in \mathbb{C}$ and $\lambda > 0$.

Proof. We adopt Theorem 1.2 to the Hilbert space $L^2(\mathbb{R})$, the unit vector $v = f$ and the position and momentum operators A and B defined above. The domain of the commutator $\mathcal{D}([A,B])$ consists precisely of the functions $f \in AC_{loc}(\mathbb{R})$ with $tf, f', tf' \in L^2(\mathbb{R})$. Moreover, for the values a and b, we take

$$a = \varepsilon_A(f) = \int_{\mathbb{R}} t|f(t)|^2 dt = t_0,$$

$$b = \varepsilon_B(f) = i \int_{\mathbb{R}} f'(t)\overline{f(t)}dt.$$

Since $[A,B]f = -if$, inequality (1.9) follows directly from inequality (1.4). Further, due to $\mathcal{F}(if')(\omega) = -2\pi\omega\mathcal{F}(f)(\omega)$ and the Parseval identity, we get

$$\varepsilon_B(f) = i \int_{\mathbb{R}} f'(t)\overline{f(t)}dt = -2\pi \int_{\mathbb{R}} \omega|\mathcal{F}(f)(\omega)|^2 d\omega = -2\pi\omega_0$$

and hence

$$\mathrm{var}_B(f) = \|if' - \varepsilon_B(f)f\|^2 = 4\pi^2 \int_{\mathbb{R}} (\omega - \omega_0)^2|\mathcal{F}(f)|^2 d\omega.$$

Therefore, inequality (1.10) is equivalent to (1.9). By Theorem 1.2, equality in (1.9) holds if and only if $if' + 2\pi\omega_0 f = -2i\lambda(t - t_0)f$ with a real constant λ. This condition implies the differential equation

$$f' = -2\lambda(t - t_0)f + 2\pi i\omega_0 f.$$

The solutions of this differential equation correspond to the Gaussian functions $G(t) = Ce^{2\pi i\omega_0 t}e^{-\lambda(t-t_0)^2}$ with constants $C \in \mathbb{C}$. Further, the constant λ has to be nonnegative in order to guarantee $G \in L^2(\mathbb{R})$. □

To keep the notation simple, we used in Theorem 1.5 the symbols tf and tf' as a shortcut for the functions given by $(tf)(t) = tf(t)$ and $(tf')(t) = tf'(t)$. We will maintain this notation in the upcoming sections. The Heisenberg-Pauli-Weyl-inequality (I.1) in the introduction is evidently also equivalent to (1.9) and (1.10).

1.3. The Breitenberger uncertainty principle

A remarkable uncertainty principle for 2π-periodic functions was formulated by Breitenberger [6] in 1983. As the Heisenberg-Pauli-Weyl principle, also the Breitenberger principle is based on the operator theoretic approach of Section 1.1.

The underlying Hilbert space of the Breitenberger principle is the space $L^2([-\pi, \pi])$ of square integrable 2π-periodic functions with inner product

$$\langle f, g \rangle := \int_{-\pi}^{\pi} f(t)\overline{g(t)}dt \qquad (1.11)$$

and norm $\|f\|^2 := \langle f, f \rangle$. On $L^2([-\pi, \pi])$, we define the operators

$$Af(t) := e^{it}f(t), \quad \mathcal{D}(A) = L^2([-\pi, \pi]), \qquad (1.12)$$

$$Bf(t) := if'(t), \quad \mathcal{D}(B) := \{f \in AC_{2\pi} : f' \in L^2([-\pi, \pi])\}. \qquad (1.13)$$

The set $AC_{2\pi}$ denotes the space of all absolutely continuous 2π-periodic functions on $[-\pi, \pi]$. For the details on absolutely continuous functions, we refer again to Section B.1. If we consider f as a L^2-density distribution on the complex unit circle, the operator A determines the angular position of the density f. In particular, the expectation value of the operator A,

$$\varepsilon(f) := \varepsilon_A(f) = \int_{-\pi}^{\pi} e^{it}|f(t)|^2 dt, \qquad (1.14)$$

can be interpreted as the center of mass (or mean value) of f in the complex plane. Further, the operator A is a unitary operator on $L^2([-\pi, \pi])$. On the other hand, the operator B gives the angular momentum of the density f. For two absolutely continuous functions $f, g \in AC_{2\pi}$, integration by parts yields

$$\int_{-\pi}^{\pi} if'(t)\overline{g(t)}dt = \int_{-\pi}^{\pi} f(t)\overline{ig'(t)}dt.$$

Thus, B is a symmetric operator and, moreover, self-adjoint (see [70, Lemma 2.3.1]). Further, the commutator of A and B is given by

$$[A, B]f = ie^{it}f' - i\frac{d}{dt}(e^{it}f) = e^{it}f = Af,$$

where $f \in \mathcal{D}([A, B]) = \mathcal{D}(B)$. Now, it is possible to prove the following uncertainty principle (cf. [6], [63], [68]):

Theorem 1.6.
Let $f \in AC_{2\pi} \subset L^2([-\pi, \pi])$ with $f' \in L^2([-\pi, \pi])$ and $\|f\| = 1$, then

$$(1 - |\varepsilon(f)|^2) \cdot \left(\|f'\|^2 - |\langle f', f\rangle|^2\right) \geq \frac{1}{4}|\varepsilon(f)|^2. \tag{1.15}$$

Equality in (1.15) is attained if and only if $f(t) = Ce^{ikt}$, $|C| = \frac{1}{\sqrt{2\pi}}$, is a normalized trigonometric monomial.

Proof. We adopt Theorem 1.4 to the Hilbert space $L^2([-\pi, \pi])$ with the unitary operator A and the symmetric operator B as defined in (1.12) and (1.13). For $\varepsilon_A(f) = \varepsilon(f)$ and $\varepsilon_B(f) = \langle if', f\rangle$, we get

$$\text{var}_A(f) = \|(A - \varepsilon_A(f))f\|^2 = 1 - |\varepsilon(f)|^2,$$
$$\text{var}_B(f) = \|(B - \varepsilon_B(f))f\|^2 = \|f'\|^2 - |\langle f', f\rangle|^2.$$

The commutator of A and B is given by $[A, B]f = ie^{it}f' - i\frac{d}{dt}(e^{it}f)$. Hence,

$$\mathcal{D}([A, B]) = \left\{f \in AC_{2\pi} : f' \in L^2([-\pi, \pi])\right\} = \mathcal{D}(B).$$

Now, inequality (1.8) implies inequality (1.15).

Due to Theorem 1.4, equality in (1.15) holds if and only if

$$if' - \varepsilon_B(f)f = \lambda(e^{it} - \varepsilon(f))f = -\bar{\lambda}(e^{-it} - \overline{\varepsilon(f)})f,$$

for a complex scalar $\lambda \in \mathbb{C}$. The second identity implies

$$f(t)\left(\lambda e^{it} + \bar{\lambda}e^{-it} - \varepsilon(f)\lambda - \overline{\lambda\varepsilon(f)}\right) = 2f(t)\left(\text{Re}(\lambda e^{-it}) - \text{Re}(\lambda\varepsilon(f))\right) = 0.$$

This condition can only be satisfied if $f = 0$ or if $\lambda = 0$. In the latter case, we get the equation $if' - \varepsilon_B(f)f = 0$. The solutions of this differential equation in $\mathcal{D}(B)$ are precisely the monomials $f(t) = Ce^{ikt}$, where $|C| = \frac{1}{\sqrt{2\pi}}$ and $\varepsilon_B(f) = k$, $k \in \mathbb{Z}$. $\qquad\square$

Motivated by Theorem 1.6, one defines

$$\text{var}_S(f) := \frac{1 - |\varepsilon(f)|^2}{|\varepsilon(f)|^2} \tag{1.16}$$

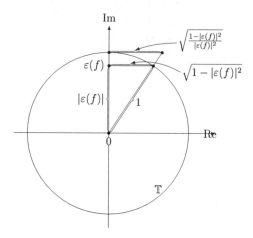

Figure 2: Geometric interpretation of the angular variance $\text{var}_S(f)$ on the complex unit circle $\mathbb{T}$. The function f is chosen such that $\varepsilon(f) = \frac{5}{6}i$ and $\text{var}_S(f) = \frac{11}{25}$.

as the angular position variance of a 2π-periodic function f (see Figure 2) and

$$\text{var}_F(f) := \|f'\|^2 - |\langle f', f \rangle|^2 \tag{1.17}$$

as the angular momentum (or frequency) variance of f. Clearly, the definition of $\text{var}_S(f)$ makes only sense if $\varepsilon(f) \neq 0$. If $\varepsilon(f) = 0$, we call the function f nowhere localized and set $\text{var}_S(f) = \infty$. Examples of nowhere localized 2π-periodic functions are the monomials $\frac{1}{\sqrt{2\pi}}e^{ikt}$, $k \in \mathbb{Z}$. Now, by Theorem 1.6, we get

Corollary 1.7 (Breitenberger uncertainty principle).
If $f \in AC_{2\pi} \subset L^2([-\pi, \pi])$ such that $f' \in L^2([-\pi, \pi])$, $\|f\| = 1$ and $\varepsilon(f) \neq 0$, then

$$\frac{1 - |\varepsilon(f)|^2}{|\varepsilon(f)|^2} \cdot (\|f'\|^2 - |\langle f', f \rangle|^2) > \frac{1}{4}. \tag{1.18}$$

Remark 1.8. In [68], Prestin and Quak showed that the constant $\frac{1}{4}$ on the right hand side of inequality (1.18) is optimal. An alternative proof for this optimality that works in a more general setting will be given in Proposition 2.58.

1.4. Uncertainty principles for weighted L^2-spaces

1.4.1. The compact case

In this section, we are going to generalize the Heisenberg-Pauli-Weyl uncertainty principle and the Breitenberger principle to the case when the Hilbert space $\mathcal{H}$ is a weighted L^2-space. In the first part, we consider as un underlying domain the interval $[0, \pi]$.

Assumption 1.9. An admissible weight function w on the interval $[0, \pi]$ satisfies the properties

$$(i) \qquad w \in AC([0, \pi]), \tag{1.19}$$

$$(ii) \qquad w(t) > 0, \quad t \in (0, \pi), \tag{1.20}$$

$$(iii) \qquad t(\pi - t)\frac{w'}{w} \in L^\infty([0, \pi]). \tag{1.21}$$

The symbol $AC([0, \pi])$ denotes the space of absolutely continuous functions f on the interval $[0, \pi]$. For a short introduction into the notion of absolute continuity, we refer again to Section B.1 of the appendix. Further, the condition (1.20) ensures that w is strictly positive in the interior $(0, \pi)$ and property (1.21) guarantees that the fraction $\frac{w'}{w}$ is not decreasing too rapidly at the possible singularity points at $t = 0$ and $t = \pi$.

Example 1.10. Consider on $[0, \pi]$ the weight function $w_{\alpha\alpha}(t) = \sin^{2\alpha+1} t$, $\alpha \geq -1/2$. Then, the properties (1.19) and (1.20) are evidently satisfied. Further,

$$\left| t(\pi - t)\frac{w'_{\alpha\alpha}(t)}{w_{\alpha\alpha}(t)} \right| = \left| t(\pi - t)(2\alpha + 1)\frac{\cos t}{\sin t} \right| \leq (2\alpha + 1)\pi.$$

Therefore, also (1.21) is satisfied. The measure $w_{\alpha\alpha}(t)dt$ corresponds to the orthogonality measure of the ultraspherical polynomials $P_l^{(\alpha,\alpha)}(\cos t)$ on $[0, \pi]$ (see Section 1.5.1).

Definition 1.11. For an admissible weight function w on $[0, \pi]$, we denote by $L^2([0, \pi], w)$ the Hilbert space of weighted square integrable functions on $[0, \pi]$ with the inner product

$$\langle f, g \rangle_w := \int_0^\pi f(t)\overline{g(t)}w(t)dt \tag{1.22}$$

and norm $\|f\|_w^2 := \langle f, f \rangle_w$.

Now, similar as in the case of the Breitenberger principle, one could think about the differential operator $B = \frac{d}{dt}$ to be a suitable momentum operator for a possible uncertainty principle on $L^2([0, \pi], w)$. However, for absolutely continuous functions $f, g \in AC([0, \pi])$, integration by parts yields

$$\int_0^\pi f'(t)g(t)w(t)dt = f(\pi)g(\pi)w(\pi) - f(0)g(0)w(0)$$

$$+ \int_0^\pi f(t)g'(t)w(t)dt + \int_0^\pi f(t)g(t)w'(t)dt.$$

Hence, the differential operator $\frac{d}{dt}$ is in general not symmetric on $L^2([0,\pi],w)$ and the two Theorems 1.2 and 1.4 can not be used to derive an uncertainty inequality in this particular case. To circumnavigate this problem, we will use a differential-difference operator T instead of the differential operator $\frac{d}{dt}$. To define such a differential-difference operator T properly, we require the symmetric extension of functions on the original interval $[0,\pi]$ to functions on the doubled interval $[-\pi,\pi]$ and the symmetric extension of the weight function w.

Definition 1.12. On the doubled interval $[-\pi,\pi]$ we define the extended weight function $\tilde{w}$ by

$$\tilde{w}(t) := \frac{1}{2}w(|t|), \quad t \in [-\pi,\pi]. \tag{1.23}$$

By $L^2([-\pi,\pi],\tilde{w})$, we denote the Hilbert space of weighted square integrable functions on $[-\pi,\pi]$ with the inner product

$$\langle f,g \rangle_{\tilde{w}} := \int_{-\pi}^{\pi} f(t)\overline{g(t)}\tilde{w}(t)dt. \tag{1.24}$$

and norm $\|f\|_{\tilde{w}}^2 := \langle f,f \rangle_{\tilde{w}}$.

Definition 1.13. On $L^2([-\pi,\pi],\tilde{w})$, we define the reflection operator $\check{}$ by

$$\check{g}(t) := g(-t), \quad \text{for a.e. } t \in [-\pi,\pi].$$

We say that a function $g \in L^2([-\pi,\pi],\tilde{w})$ is even (odd) if $\check{g} = g$ ($\check{g} = -g$, respectively). The subspace of even functions in $L^2([-\pi,\pi],\tilde{w})$ is denoted by

$$L_e^2([-\pi,\pi],\tilde{w}) := \left\{ g \in L^2([-\pi,\pi],\tilde{w}) : \check{g} = g \right\}. \tag{1.25}$$

The notion 'for a.e. $t \in [-\pi,\pi]$' in Definition 1.13 means that the statement holds for all $t \in [-\pi,\pi]$ except a set of Lebesgue measure zero. On $L^2([0,\pi],w)$ and $L^2([-\pi,\pi],\tilde{w})$, we can introduce the even extension operator e and the restriction operator r as

$$e : L^2([0,\pi],w) \to L^2([-\pi,\pi],\tilde{w}), \quad e(f)(t) := f(|t|), \tag{1.26}$$
$$r : L^2([-\pi,\pi],\tilde{w}) \to L^2([0,\pi],w), \quad r(g) := g|_{[0,\pi]}. \tag{1.27}$$

Then, the operators e and r constitute isometric isomorphisms between $L^2([0,\pi],w)$ and the space of even functions $L_e^2([-\pi,\pi],\tilde{w})$.

Since $w \in AC([0,\pi])$, its symmetrically extended function $\tilde{w} \in AC_{2\pi}$ is absolutely continuous and 2π-periodic. In particular, its Radon-Nikodym derivative, denoted as

$$\tilde{w}' = \frac{d\tilde{w}}{dt}, \tag{1.28}$$

is an element of $L^1([-\pi,\pi])$.

Definition 1.14. For an admissible weight function w satisfying Assumption 1.9, we define on $L^2([-\pi,\pi],\tilde{w})$ the Dunkl operator T by

$$Tg := \frac{dg}{dt} + \frac{\tilde{w}'}{\tilde{w}}\frac{g-\check{g}}{2}, \qquad (1.29)$$

with the domain

$$\mathcal{D}(T) := \left\{ g \in L^2([-\pi,\pi],\tilde{w}) \cap AC_{2\pi} : \frac{dg}{dt}, \frac{\tilde{w}'}{\tilde{w}}\frac{g-\check{g}}{2} \in L^2([-\pi,\pi],\tilde{w}) \right\}. \qquad (1.30)$$

Remark 1.15. The fact that w satisfies Assumption 1.9 plays a very important role in Definition 1.14. It guarantees that the domain $\mathcal{D}(T)$ of the Dunkl operator T is dense in the Hilbert space $L^2([-\pi,\pi],\tilde{w})$. In fact, we can show that the trigonometric polynomials form a subset of $\mathcal{D}(T)$. Obviously, if $\mathcal{P}(t) = \sum_{l=-n}^{n} c_l e^{ilt}$ is a trigonometric polynomial, then $\mathcal{P}$ is absolutely continuous on $[-\pi,\pi]$, $\mathcal{P}(-\pi) = \mathcal{P}(\pi)$ and $\mathcal{P}' \in L^2([-\pi,\pi],\tilde{w})$. Further, we have

$$\int_{-\pi}^{\pi} \left| \frac{\tilde{w}'(t)}{\tilde{w}(t)} \frac{\mathcal{P}(t) - \mathcal{P}(-t)}{2} \right|^2 \tilde{w}(t)dt$$

$$= \int_{-\pi}^{\pi} \left| t(\pi^2 - t^2) \frac{\tilde{w}'(t)}{\tilde{w}(t)} \right|^2 \left| \sum_{l=1}^{n} \frac{c_l - c_{-l}}{2} \frac{\sin(lt)}{t(\pi^2 - t^2)} \right|^2 \tilde{w}(t)dt$$

$$\leq 2 \left\| t(\pi - t)\frac{w'}{w} \right\|_{\infty}^2 \left(\sum_{l=1}^{n} |c_l - c_{-l}| \left\| \frac{\sin(lt)}{t(\pi - t)} \right\|_w \right)^2.$$

Hence, since the weight function w satisfies property (1.21) and $\frac{\sin(lt)}{t(\pi-t)} \in L^2([0,\pi],w)$, we conclude that also $\frac{\tilde{w}'}{\tilde{w}}\frac{\mathcal{P}-\check{\mathcal{P}}}{2} \in L^2([-\pi,\pi],\tilde{w})$.

Whereas the differential operator $\frac{d}{dt}$ is in general not symmetric on $L^2([0,\pi],w)$, we can now show that the Dunkl operator iT is symmetric on the extended Hilbert space $L^2([-\pi,\pi],\tilde{w})$. Precisely this Dunkl operator will be used afterwards to determine an uncertainty principle for functions in $L^2([0,\pi],w)$.

Lemma 1.16.
The operator iT is symmetric and densely defined on $L^2([-\pi,\pi],\tilde{w})$. Moreover, if $\tilde{w}$ is strictly positive on the whole interval $[-\pi,\pi]$, then iT is self-adjoint.

Proof. By the Stone-Weierstrass Theorem B.1, the trigonometric polynomials form a dense subspace of $C_{2\pi}$, and thus also a dense subspace of $L^2([-\pi,\pi],\tilde{w})$. Hence, by Remark 1.15, also $\mathcal{D}(T)$ is a dense subset of $L^2([-\pi,\pi],\tilde{w})$.
To prove the symmetry of iT, we follow in principle the proof of [73, Lemma 3.1]. For two absolutely continuous functions, integration by parts is well defined (see equation (B.4) in the appendix). So, for $f,g \in \mathcal{D}(T)$, we get

$$\int_{-\pi}^{\pi} f'(t)\overline{g(t)}\tilde{w}(t)dt = -\int_{-\pi}^{\pi} f(t)\frac{d}{dt}\left(\overline{g(t)}\tilde{w}(t) \right)dt$$

$$= -\int_{-\pi}^{\pi} f(t)\left(\overline{g'(t)} + \overline{g(t)}\frac{\tilde{w}'(t)}{\tilde{w}(t)} \right)\tilde{w}(t)dt.$$

Now, by definition of the operator T, we get

$$
\int_{-\pi}^{\pi} (iTf)(t)\overline{g(t)}\tilde{w}(t)dt = -i\int_{-\pi}^{\pi}\left(f(t)\overline{g'(t)} + f(t)\overline{g(t)}\frac{\overline{\tilde{w}'(t)}}{\overline{\tilde{w}(t)}}\right)\tilde{w}(t)dt
$$
$$
+ i\int_{-\pi}^{\pi}\frac{f(t)-f(-t)}{2}\overline{g(t)}\frac{\tilde{w}'(t)}{\tilde{w}(t)}\tilde{w}(t)dt
$$
$$
= -i\int_{-\pi}^{\pi}\left(f(t)\overline{g'(t)} + \frac{f(t)+f(-t)}{2}\overline{g(t)}\frac{\overline{\tilde{w}'(t)}}{\overline{\tilde{w}(t)}}\right)\tilde{w}(t)dt
$$
$$
= -i\int_{-\pi}^{\pi}\left(f(t)\overline{g'(t)} + f(t)\overline{\frac{g(t)-g(-t)}{2}}\frac{\tilde{w}'(t)}{\tilde{w}(t)}\right)\tilde{w}(t)dt
$$
$$
= \int_{-\pi}^{\pi} f(t)\overline{(iTg)(t)}\tilde{w}(t)dt.
$$

Therefore, the operator iT is symmetric. In particular, the domain $\mathcal{D}(T)$ is a subset of the domain $\mathcal{D}(T^*)$ of the dual operator T^* (see Section B.3).

We show now that iT is self-adjoint if $\tilde{w}$ is strictly positive on the whole interval $[-\pi,\pi]$. This can be done similarly as in [76, Example 13.4] by proving that also the inclusion $\mathcal{D}(T^*) \subset \mathcal{D}(T)$ holds. Therefore, we take a function $g \in \mathcal{D}(T^*)$ and set $f = T^*g \in L^2([-\pi,\pi],\tilde{w})$. Next, we define as a generalized primitive of f (with respect to the Dunkl operator T), the function

$$
F(t) = \frac{1}{\tilde{w}(t)}\int_{-\pi}^{t}\frac{f(\tau)+f(-\tau)}{2}\tilde{w}(\tau)d\tau + \int_{-\pi}^{t}\frac{f(\tau)-f(-\tau)}{2}d\tau. \tag{1.31}
$$

Since $f \in L^2([-\pi,\pi],\tilde{w}) \subset L^1([-\pi,\pi])$, F is well defined and absolutely continuous on $[-\pi,\pi]$ by the fundamental theorem of calculus for the Lebesgue integral. Moreover, we get for a.e. $t \in [-\pi,\pi]$

$$
TF(t) = \frac{f(t)+f(-t)}{2} - \frac{\tilde{w}'(t)}{\tilde{w}(t)^2}\int_{-\pi}^{t}\frac{f(\tau)+f(-\tau)}{2}\tilde{w}(\tau)d\tau
$$
$$
+ \frac{\tilde{w}'(t)}{\tilde{w}(t)^2}\int_{-\pi}^{t}\frac{f(\tau)+f(-\tau)}{2}\tilde{w}(\tau)d\tau + \frac{f(t)-f(-t)}{2} = f(t),
$$

and for $F(\pi)$

$$
F(\pi) = \frac{1}{\tilde{w}(\pi)}\int_{-\pi}^{\pi}\frac{f(\tau)+f(-\tau)}{2}\tilde{w}(\tau)d\tau = \frac{1}{\tilde{w}(\pi)}\left\langle\frac{f+\check{f}}{2},1\right\rangle_{\tilde{w}}
$$
$$
= \frac{1}{\tilde{w}(\pi)}\left\langle\frac{T^*g+T^*\check{g}}{2},1\right\rangle_{\tilde{w}} = \frac{1}{\tilde{w}(\pi)}\left\langle\frac{g+\check{g}}{2},T1\right\rangle_{\tilde{w}} = 0.
$$

Hence, F is 2π-periodic and, in particular, an element of $\mathcal{D}(T)$. Now, using the symmetry of the operator T, we get for all $\varphi \in \mathcal{D}(T)$

$$
\langle T\varphi,g\rangle_{\tilde{w}} = \langle\varphi,T^*g\rangle_{\tilde{w}} = \langle\varphi,f\rangle_{\tilde{w}} = \langle\varphi,TF\rangle_{\tilde{w}} = -\langle T\varphi,F\rangle_{\tilde{w}}. \tag{1.32}
$$

For $f \in L^2([-\pi, \pi], \tilde{w})$, the generalized primitive F in (1.31) is an element of $\mathcal{D}(T)$ if and and only if $F(\pi) = 0$. Hence, the range of the Dunkl operator T is the set of all $f \in L^2([-\pi, \pi], \tilde{w})$ for which $\int_{-\pi}^{\pi} f(\tau)\tilde{w}(\tau)d\tau = 0$ holds, i.e. range$(T)^{\perp}$ is the subspace of constant functions in $L^2([-\pi, \pi], \tilde{w})$. Thus, equation (1.32) implies that $g = -F + C$ for a constant $C \in \mathbb{C}$. Since $-F + C$ is absolutely continuous and 2π-periodic and $g' \in L^2([-\pi, \pi], \tilde{w})$, it follows that $g \in \mathcal{D}(T)$ and, hence, that iT is self-adjoint. $\qquad\square$

Example 1.17. Consider the weight function $w(t) = t^4$ on $[0, \pi]$. Then, the Dunkl operator on $L^2([-\pi, \pi], \tilde{w})$ attains the form

$$Tg(t) = \frac{dg}{dt}(t) + 2\frac{g(t) - g(-t)}{t}.$$

The function $h \in L^2([-\pi, \pi], \tilde{w})$ given by $h(t) = |t|^{-1}$ is not absolutely continuous on $[-\pi, \pi]$ and therefore not in $\mathcal{D}(T)$. On the other hand, for any $g \in \mathcal{D}(T)$ integration by parts yields

$$|\langle Tg, h\rangle_{\tilde{w}}| = \left| \int_{-\pi}^{\pi} \left(\frac{dg}{dt} + \frac{2}{t}(g(t) - g(-t)) \right) |t|^3 dt \right|$$
$$= \left| \int_{-\pi}^{\pi} (g(t) + 2g(-t)) \operatorname{sign}(t)|t|^2 dt \right| \leq 3 \left| \int_{-\pi}^{\pi} |g(t)||t|^2 dt \right| \leq 6\pi \|g\|_{\tilde{w}}.$$

Hence, the functional $\langle Tg, h\rangle_{\tilde{w}}$ is continuous on $\mathcal{D}(T)$. This implies that h is an element of $\mathcal{D}(T^*)$ and that $\mathcal{D}(T) \subsetneq \mathcal{D}(T^*)$. So, if the weight function w is not strictly positive on the whole interval $[0, \pi]$, the operator iT is in general not self-adjoint.

Using Theorem 1.4 and the Dunkl operator T, we can deduce an uncertainty principle for functions in the extended Hilbert space $L^2([-\pi, \pi], \tilde{w})$. We fix a function $h \in \mathcal{D}(T)$ and define two operators A and B on $L^2(X^d, \tilde{W})$ by

$$Ag = hg, \qquad \mathcal{D}(A) = L^2([-\pi, \pi], \tilde{w}), \tag{1.33}$$
$$Bg = iTg, \qquad \mathcal{D}(B) = \mathcal{D}(T). \tag{1.34}$$

The multiplication operator A is a normal and bounded operator on $L^2([-\pi, \pi], \tilde{w})$. The differential-difference operator B is symmetric due to Lemma 1.16. So, the next result is an immediate consequence of Theorem 1.4.

Theorem 1.18.
For a fixed multiplier $h \in \mathcal{D}(T)$ and an even function $g \in \mathcal{D}(T) \cap L_e^2([-\pi, \pi], \tilde{w})$, the following inequality holds:

$$\left(\|hg\|_{\tilde{w}}^2 - \frac{|\langle hg, g\rangle_{\tilde{w}}|^2}{\|g\|_{\tilde{w}}} \right) \cdot \|g'\|_{\tilde{w}}^2 \geq \frac{1}{4}|\langle g\,Th, g\rangle_{\tilde{w}}|. \tag{1.35}$$

Proof. We consider the operators A and B defined in (1.33) and (1.34). For an even function $g \in \mathcal{D}(T)$, the derivative $Tg = g'$ is odd. Thus, $\langle Bg, g\rangle_{\tilde{w}} - \langle ig', g\rangle_w = 0$ and

$\langle Ag, g \rangle_{\tilde{w}} = \langle hg, g \rangle_{\tilde{w}}$. The commutator $[A, B]$ of the operator A and B acting on functions $g \in \mathcal{D}(AB) \cap \mathcal{D}(BA) = \mathcal{D}(T)$ is given by

$$[A, B]g(t) = ihTg(t) - iT(hg)(t) = -i\left(h'(t)g(t) + \frac{w'(t)}{w(t)}\frac{h(t) - h(-t)}{2}\right)g(-t),$$

for a.e. $t \in [-\pi, \pi]$. Further, since $g \in L_e^2([-\pi, \pi], \tilde{w}) \cap \mathcal{D}(T)$ is even, we get

$$[A, B]g = -(iTh)g. \tag{1.36}$$

Inequality (1.35) now follows from inequality (1.8) with $a = \langle hg, g \rangle_{\tilde{w}}$, $b = \langle Bg, g \rangle_{\tilde{w}} = 0$ and formula (1.36) for the commutator. $\quad\square$

Theorem 1.18 gives a natural way to formulate an uncertainty principle for the initial Hilbert space $L^2([0, \pi], w)$. Namely, if $f \in L^2([0, \pi], w)$, we can take the even extension $e(f) \in L_e^2([-\pi, \pi], \tilde{w})$ and then use inequality (1.35). A common choice for the multiplier h in the definition (1.33) of the operator A is the 2π-periodic function $h(t) = e^{it}$ (see [27, (2.13)], [73, Theorem 2.2] and [79, Theorem 9.2.]). With this particular choice for the multiplier h, we get

Theorem 1.19.
Assume that the weight function w satisfies Assumption 1.9. Let $f \in L^2([0, \pi], w) \cap AC([0, \pi])$ with $f' \in L^2([0, \pi], w)$ and normalized such that $\|f\|_w = 1$. Then, the following uncertainty principle holds:

$$\left(1 - \left(\int_0^\pi \cos t\, |f(t)|^2 w(t)dt\right)^2\right) \cdot \|f'\|_w^2 \geq$$
$$\frac{1}{4}\left|\int_0^\pi \left(\cos t\, w(t) + \sin t\, w'(t)\right)|f(t)|^2 dt\right|^2. \tag{1.37}$$

Equality in (1.37) can only be attained if f is a constant function.

Proof. If $f \in AC([0, \pi])$ with $f' \in L^2([0, \pi], w)$ and $\|f\|_w = 1$, then $e(f) \in \mathcal{D}(T)$ and $\|e(f)\|_{\tilde{w}} = 1$. Now, we can adopt inequality (1.35) to prove (1.37). As a multiplier h, we choose $h(t) = e^{it}$, $t \in [-\pi, \pi]$. Then $h \in \mathcal{D}(T)$, and we get

$$\|he(f)\|_{\tilde{w}}^2 = \|e(f)\|_{\tilde{w}}^2 = \|f\|_w^2 = 1,$$
$$|\langle he(f), e(f) \rangle_{\tilde{w}}|^2 = |\langle e^{it}e(f), e(f) \rangle_{\tilde{w}}|^2 = \left(\int_0^\pi \cos t |f(t)|^2 w(t)dt\right)^2,$$
$$\|e(f)'\|_{\tilde{w}} = \|f'\|_w,$$
$$iTh(t) = \left(-e^{it} - \frac{\tilde{w}'(t)}{\tilde{w}(t)}\sin t\right), \quad \text{for a.e. } t \in [-\pi, \pi].$$

Further, since $\sin t$ and $\frac{\tilde{w}'}{\tilde{w}}$ are odd functions on $[-\pi, \pi]$, we conclude

$$\langle (iTh)e(f), e(f) \rangle_{\tilde{w}} = \int_{-\pi}^{\pi} \left(-e^{it} - \frac{\tilde{w}'(t)}{\tilde{w}(t)} \sin t \right) |e(f)(t)|^2 \tilde{w}(t) dt$$

$$= -\int_{0}^{\pi} \left(\cos t\, w(t) + \sin t\, w'(t) \right) |f(t)|^2 dt.$$

Hence, inequality (1.37) holds. Due to Theorem 1.8, equality in (1.35) is attained if and only if (note that $a = \langle e^{it}e(f), e(f) \rangle_{\tilde{w}}$ and $b = 0$)

$$ie(f)' = \lambda(e^{it} - a)f = -\bar{\lambda}(e^{-it} - \bar{a})f, \quad \lambda \in \mathbb{C}.$$

This identity implies

$$f(t)\left(\lambda e^{it} + \bar{\lambda}e^{-it} - a\lambda - \bar{\lambda}\bar{a} \right) = 2f(t)\left(\operatorname{Re}(\lambda e^{-it}) - \operatorname{Re}(a\lambda) \right) = 0, \quad t \in [-\pi, \pi].$$

This condition can only be satisfied if $f = 0$ or if $\lambda = 0$. In the latter case we get $if' = 0$. Thus, f has to be constant on $[0, \pi]$. $\qquad \square$

Similar to the Breitenberger uncertainty principle (1.18), we can introduce a generalized mean value $\varepsilon(f)$ for a function $f \in L^2([0, \pi], w)$ by

$$\varepsilon(f) := \langle e^{it}e(f), e(f) \rangle_{\tilde{w}} = \int_{0}^{\pi} \cos t\, |f(t)|^2 w(t) dt. \qquad (1.38)$$

Moreover, we denote the integral term on the right hand side of (1.37) as

$$\rho(f) := \langle -(iTe^{it})e(f), e(f) \rangle_{\tilde{w}} = \int_{0}^{\pi} \left(\cos t\, w(t) + \sin t\, w'(t) \right) |f(t)|^2 dt. \qquad (1.39)$$

Definition 1.20. If $\rho(f) \neq 0$, we define

$$\operatorname{var}_S(f) := \frac{1 - \varepsilon(f)^2}{\rho(f)^2}, \qquad (1.40)$$

$$\operatorname{var}_F(f) := \|f'\|_w. \qquad (1.41)$$

The values $\operatorname{var}_S(f)$ and $\operatorname{var}_F(f)$ are called the position and the frequency variance of f, respectively.

Corollary 1.21.
Let w satisfy Assumption 1.9 and let $f \in L^2([0, \pi], w) \cap AC([0, \pi])$ with $f' \in L^2([0, \pi], w)$, $\|f\|_w = 1$ and $\rho(f) \neq 0$. Then, the following uncertainty principle holds:

$$\operatorname{var}_S(f) \cdot \operatorname{var}_F(f) > \frac{1}{4}. \qquad (1.42)$$

Proof. Clearly, (1.42) follows from (1.37). It remains to check the strict inequality in (1.42). The only functions for which equality is attained in (1.37) are the constant functions. If $f = C$ is constant on $[0, \pi]$, integration by parts with respect to the variable t yields

$$\rho(f) = |C| \int_0^\pi \Big(\cos t \, w(t) + \sin t \, w'(t) \Big) dt = 0.$$

Hence, there exists no function f with $f \in L^2([0, \pi], w) \cap AC([0, \pi])$, $f' \in L^2([0, \pi], w)$, $\|f\|_w = 1$ and $\rho(f) \neq 0$ for which equality holds in (1.42). $\qquad \square$

1.4.2. The compact case with zero boundary condition

In Theorem 1.18, the multiplier h has to be an element of the domain $\mathcal{D}(T)$ in order that the commutator $[A, B]$ is well defined for functions in $\mathcal{D}(T)$. This is certainly the case if h is given by $h(t) = e^{it}$, but not if we choose, for instance, $h(t) = t$, $t \in [-\pi, \pi]$. However, the option $h(t) = t$ is possible, if we restrict the domain of the Dunkl operator T to functions $g \in AC([-\pi, \pi])$ satisfying the boundary condition $g(-\pi) = g(\pi) = 0$.

Definition 1.22. Consider the Hilbert space $L^2([0, \pi], w)$ and the extension $L^2([-\pi, \pi], \tilde{w})$ as in Definition 1.12. Then, we restrict the Dunkl operator T defined in (1.29) by $Tg = g' + \frac{\tilde{w}'}{\tilde{w}} \frac{g - \check{g}}{2}$ to the smaller domain

$$\mathcal{D}_0(T) := \Big\{ g \in AC_{2\pi} : g(\pi) = 0, \; \frac{dg}{dt}, \; \frac{\tilde{w}'}{\tilde{w}} \frac{g - \check{g}}{2}, \; t \frac{\tilde{w}'}{\tilde{w}} g \in L^2([-\pi, \pi], \tilde{w}) \Big\}. \qquad (1.43)$$

In view of property (1.21) of the weight function w, the condition $t \frac{\tilde{w}'}{\tilde{w}} g \in L^2([-\pi, \pi], \tilde{w})$ in (1.43) is a growth condition on the function g at the point $t = \pi$. This additional condition will be needed in the proof of Theorem 1.23 below. Since $\mathcal{D}_0(T) \subset \mathcal{D}(T)$, Lemma 1.16 implies that the operator iT is also symmetric on the smaller domain $\mathcal{D}_0(T)$.

On $L^2([-\pi, \pi], \tilde{w})$, we consider now the bounded operator A defined by $Ag = hg$ and the operator $B = iT$ defined on the smaller domain $\mathcal{D}_0(T)$. In the following, we will show that the commutator $[A, B]$ is well defined for functions in $\mathcal{D}_0(T)$ if the multiplier h is an element of

$$\mathcal{M} := \Big\{ h \in AC([-\pi, \pi]) \cap L^2([-\pi, \pi], \tilde{w}) : \; \frac{dh}{dt} \in L^2([-\pi, \pi], \tilde{w}),$$

$$(\pi^2 - t^2) \frac{\tilde{w}'}{\tilde{w}} \frac{h - \check{h}}{2} \in L^2([-\pi, \pi], \tilde{w}) \Big\}. \qquad (1.44)$$

Note that the functions $h \in \mathcal{M}$ are not supposed to fulfill the periodicity condition $h(-\pi) = h(\pi)$ and that the condition $(\pi^2 - t^2) \frac{\tilde{w}'}{\tilde{w}} \frac{h - \check{h}}{2} \in L^2([-\pi, \pi], \tilde{w})$ is weaker than the condition $\frac{\tilde{w}'}{\tilde{w}} \frac{h - \check{h}}{2} \in L^2([-\pi, \pi], \tilde{w})$ in the domain $\mathcal{D}(T)$. Now, similar to Theorem 1.18, we get the following result for even functions in $\mathcal{D}_0(T)$:

Theorem 1.23.
For a fixed $h \in \mathcal{M}$ and an even function $g \in \mathcal{D}_0(T) \cap L_e^2([-\pi, \pi], \tilde{w})$, we have the inequality

$$\left(\|hg\|_{\tilde{w}}^2 - \frac{|\langle hg, g \rangle_{\tilde{w}}|^2}{\|g\|_{\tilde{w}}} \right) \cdot \|g'\|_{\tilde{w}}^2 \geq \frac{1}{4} |\langle hTg - T(hg), g \rangle_{\tilde{w}}|. \tag{1.45}$$

Proof. The operator A defined by $Ag = hg$, $h \in \mathcal{M}$ is a normal and bounded operator on $L^2([-\pi, \pi], \tilde{w})$ and the operator $B = iT$ with domain $\mathcal{D}(B) = \mathcal{D}_0(T)$ is symmetric due to Lemma 1.16. For $g \in \mathcal{D}_0(T)$ and $h \in \mathcal{M}$, we have that $hg \in AC([-\pi, \pi])$, $hg(-\pi) = hg(\pi) = 0$ and that $\frac{d(hg)}{dt} \in L^2(X^d, \tilde{W})$. Further, we have

$$\left\| \frac{\tilde{w}'}{\tilde{w}} \frac{hg - \check{h}\check{g}}{2} \right\|_{\tilde{w}}^2 = \left\| \frac{\tilde{w}'}{\tilde{w}} \frac{h - \check{h}}{2} \frac{g + \check{g}}{2} + \frac{\tilde{w}'}{\tilde{w}} \frac{g - \check{g}}{2} \frac{h + \check{h}}{2} \right\|_{\tilde{w}}^2$$

$$\leq \left\| (\pi^2 - t^2 + |t|) \frac{\tilde{w}'}{\tilde{w}} \frac{h - \check{h}}{2} \frac{g + \check{g}}{2} \right\|_{\tilde{w}}^2 + \left\| \frac{\tilde{w}'}{\tilde{w}} \frac{g - \check{g}}{2} \right\|_{\tilde{w}}^2 \|h\|_\infty$$

$$\leq \left\| (\pi^2 - t^2) \frac{\tilde{w}'}{\tilde{w}} \frac{h - \check{h}}{2} \right\|_{\tilde{w}}^2 \|g\|_\infty + \left\| t \frac{\tilde{w}'}{\tilde{w}} g \right\|_{\tilde{w}}^2 \|h\|_\infty + \left\| \frac{\tilde{w}'}{\tilde{w}} \frac{g - \check{g}}{2} \right\|_{\tilde{w}}^2 \|h\|_\infty.$$

Hence, by definition of the domain $\mathcal{D}_0(T)$ and the set $\mathcal{M}$, the product hg is in $\mathcal{D}_0(T)$ and the commutator $[A, B]g = h(iTg) - iT(hg)$ is well defined for functions $g \in \mathcal{D}_0(T^X)$. Inequality (1.45) now follows from inequality (1.8) with $a = \langle hg, g \rangle_{\tilde{w}}$ and $b = \langle iTg, g \rangle_{\tilde{w}} = \langle ig', g \rangle_{\tilde{w}} = 0$. □

Since $\mathcal{M} \supset \mathcal{D}(T)$, the set of admissible multipliers h in Theorem 1.23 is larger than in Theorem 1.18. In particular, the multiplier h given by $h(t) = t$ is an element of $\mathcal{M} \backslash \mathcal{D}(T)$.

Theorem 1.24.
Assume that the weight function w satisfies Assumption 1.9. Let $f \in L^2([0, \pi], w) \cap AC([0, \pi])$, $f' \in L^2([0, \pi], w)$, satisfying the zero boundary condition $f(\pi) = 0$ and $t \frac{w'}{w} f \in L^2([0, \pi], w)$. Further, let f be normalized such that $\|f\|_w = 1$. Then, the following uncertainty inequality holds:

$$\|tf\|_w^2 \cdot \|f'\|_w^2 > \frac{1}{4} \left| 1 + \int_0^\pi tw'(t)|f(t)|^2 dt \right|^2. \tag{1.46}$$

Proof. We proceed as in the proof of Theorem 1.19. If $f \in AC([0, \pi])$ with $f' \in L^2([0, \pi])$, zero boundary condition $f(\pi) = 0$ and $t \frac{w'}{w} f \in L^2([0, \pi], w)$, then $e(f) \in \mathcal{D}_0(T)$. Now, inserting $g = e(f)$ and $h(t) = t$ in inequality (1.45), we get

$$\|he(f)\|_{\tilde{w}}^2 = \|te(f)\|_{\tilde{w}}^2 = \|tf\|_w^2,$$

$$\langle he(f), e(f) \rangle_{\tilde{w}} = \langle te(f), e(f) \rangle_{\tilde{w}} = 0,$$

$$\|e(f)'\|_{\tilde{w}} = \|f'\|_w,$$

$$(T(he(f)) - hTe(f))(t) = e(f)(t) + \frac{\tilde{w}'(t)}{\tilde{w}(t)} te(f)(t), \quad \text{for a.e. } t \in [-\pi, \pi].$$

25

Further, since t and $\frac{\tilde{w}'}{\tilde{w}}$ are odd functions on $[-\pi, \pi]$, we conclude

$$\langle T(he(f)) - hTe(f), e(f)\rangle_{\tilde{w}} = \int_{-\pi}^{\pi} \left(1 + \frac{\tilde{w}'(t)}{\tilde{w}(t)}t\right)|e(f)(t)|^2 \tilde{w}(t)dt$$
$$= 1 + \int_{0}^{\pi} tw'(t)|f(t)|^2 dt.$$

The operator A given by $Ag = tg$ is symmetric. Since also $B = iT$ is symmetric, Theorem 1.2 states that equality in (1.45) holds if and only if (note that $a = \langle te(f), e(f)\rangle_{\tilde{w}} = 0$, $b = \langle ie(f)', e(f)\rangle_{\tilde{w}} = 0$)

$$ie(f)'(t) = i\lambda te(f)(t), \quad t \in [-\pi, \pi], \ \lambda \in \mathbb{R},$$

i.e. if and only if $f(t) = Ce^{-\lambda t^2}$. Since f has to satisfy the boundary condition $f(\pi) = 0$, this can only hold for the zero function. But the zero function is not a permissible function in Theorem 1.24. Hence, inequality (1.46) is strict. $\qquad \square$

1.4.3. The non-compact case

In this last part, we consider weighted square integrable functions on the nonnegative real half axis $[0, \infty)$. The resulting uncertainty principle will be a generalization of the Heisenberg-Pauli-Weyl inequality. Hereby, the weight function w has to satisfy the following conditions:

Assumption 1.25. An admissible weight functions w on the nonnegative real half axis $[0, \infty)$ satisfies the properties

$$(i) \qquad w \in AC_{loc}([0, \infty)), \tag{1.47}$$
$$(ii) \qquad w(t) > 0, \quad t \in (0, \infty), \tag{1.48}$$
$$(iii) \qquad t\frac{w'}{w}\Big|_{[a,b]} \in L^{\infty}([a, b]), \quad \text{for all } [a, b] \subset [0, \infty). \tag{1.49}$$

Assumption 1.25 resembles Assumption 1.9 on the interval $[0, \pi]$. The conditions (1.47) and (1.48) guarantee that w is strictly positive on $(0, \infty)$ and locally absolutely continuous. The property (1.49) ensures that the fraction $t\frac{w'}{w}$ is essentially bounded on every interval $[a, b] \subset [0, \infty)$. Further, there are no integrability restrictions on the weight function w and, in particular, the integral $\int_0^\infty w(t)dt$ is not supposed to be finite. This is, for instance, the case in the Heisenberg-Pauli-Weyl principle (Theorem 1.5), where $w(t) = 1$.

Example 1.26. Consider the weight function $w_\alpha(t) = t^\alpha e^{-t}$, $\alpha \geq 0$, on $[0, \infty)$. The conditions (1.47) and (1.48) are evidently satisfied. Further,

$$\left|t\frac{w'_\alpha(t)}{w_\alpha(t)}\right| = t\left|\frac{\alpha}{t} - 1\right| = |\alpha - t|.$$

Hence, also condition (1.49) is satisfied. The weight function w_α determines the orthogonality measure of the Laguerre polynomials (see Section 1.5.2).

Definition 1.27. For an admissible weight function w on $[0, \infty)$, we denote by $L^2([0, \infty), w)$ the Hilbert space of weighted square integrable functions on $[0, \infty)$ with the inner product

$$\langle f, g \rangle_w := \int_0^\infty f(t)\overline{g(t)}w(t)dt \qquad (1.50)$$

and norm $\|f\|_w^2 := \langle f, f \rangle_w$.

Similar as in Definition 1.12, we can also define a symmetric extended weight function $\tilde{w}$ on the real axis $\mathbb{R}$.

Definition 1.28. On $\mathbb{R}$, we define the symmetric extended weight function $\tilde{w}$ by

$$\tilde{w}(t) := \frac{1}{2}w(|t|), \quad t \in \mathbb{R}. \qquad (1.51)$$

and denote by $L^2(\mathbb{R}, \tilde{w})$ the Hilbert space of weighted square integrable functions on $\mathbb{R}$ with the inner product

$$\langle f, g \rangle_{\tilde{w}} := \int_{\mathbb{R}} f(t)\overline{g(t)}\tilde{w}(t)dt \qquad (1.52)$$

and norm $\|f\|_{\tilde{w}}^2 := \langle f, f \rangle_{\tilde{w}}$.

To relate the Hilbert spaces $L^2([0, \infty), w)$ and $L^2(\mathbb{R}, \tilde{w})$, we introduce the even extension operator e and the restriction operator r as

$$e : L^2([0, \infty), w) \to L^2(\mathbb{R}, \tilde{w}), \quad e(f)(t) := f(|t|), \qquad (1.53)$$
$$r : L^2(\mathbb{R}, \tilde{w}) \to L^2([0, \infty), w), \quad r(g) := g|_{[0,\infty)}. \qquad (1.54)$$

The operators e and r define isometric isomorphisms between $L^2([0, \infty), w)$ and the space of even functions

$$L_e^2(\mathbb{R}, \tilde{w}) := \left\{ g \in L^2(\mathbb{R}, \tilde{w}) : \check{g} = g \right\}.$$

Hereby, $\check{g}$ is defined as in the compact case by $\check{g}(t) = g(-t)$ for a.e. $t \in \mathbb{R}$.

Since $w \in AC_{loc}([0, \infty))$, the even function $\tilde{w} \in AC_{loc}(\mathbb{R})$ is locally absolutely continuous on $\mathbb{R}$. Its Radon-Nikodym derivative, denoted as $\tilde{w}' = \frac{d\tilde{w}}{dt}$, is locally integrable, i.e., $\tilde{w}'|_{[a,b]} \in L^1([a, b])$ for all intervals $[a, b] \in \mathbb{R}$.

Definition 1.29. For a weight function w satisfying Assumption 1.25, we define on $L^2(\mathbb{R}, \tilde{w})$ the Dunkl operator T by

$$Tg := \frac{dg}{dt} + \frac{\tilde{w}'}{\tilde{w}} \frac{g - \check{g}}{2} \qquad (1.55)$$

with the domain

$$\mathcal{D}(T) = \left\{ g \in AC_{loc}(\mathbb{R}) \cap L^2(\mathbb{R}, \tilde{w}) : \frac{dg}{dt}, \frac{\tilde{w}'}{\tilde{w}} \frac{g - \check{g}}{2} \in L^2(\mathbb{R}, \tilde{w}) \right\}. \qquad (1.56)$$

Remark 1.30. The fact that w satisfies Assumption 1.25 is crucial in Definition 1.29. It ensures that the domain $\mathcal{D}(T)$ of the Dunkl operator T is dense in $L^2(\mathbb{R}, \tilde{w})$. Namely, we can show that the C^∞-functions with compact support in $\mathbb{R}$, denoted as $C_c^\infty(\mathbb{R})$, are included in $\mathcal{D}(T)$. Clearly, if $g \in C_c^\infty(\mathbb{R})$, then g is locally absolutely continuous on $\mathbb{R}$ and $g' \in C_c^\infty(\mathbb{R}) \subset L^2(\mathbb{R}, \tilde{w})$. Further, if $\mathrm{supp}(g) \subset [-K, K]$, $K > 0$, we have

$$\int_{\mathbb{R}} \left| \frac{\tilde{w}'(t)}{\tilde{w}(t)} \frac{g(t) - g(-t)}{2} \right|^2 \tilde{w}(t) dt$$

$$= \int_{-K}^{K} \left| t \frac{\tilde{w}'(t)}{\tilde{w}(t)} \right|^2 \left| \frac{g(t) - g(-t)}{2t} \right|^2 \tilde{w}(t) dt \leq \left\| t \frac{w'}{w} \right\|_{L^\infty([0,K])}^2 \left\| \frac{g - \check{g}}{2t} \right\|_{\tilde{w}}^2.$$

Hence, since w satisfies the property (1.49) and $\frac{g-\check{g}}{t} \in L^2(\mathbb{R}, \tilde{w})$, we conclude that also $\frac{\tilde{w}'}{\tilde{w}} \frac{g-\check{g}}{2} \in L^2(\mathbb{R}, \tilde{w})$.

Now, as in the compact case (see Lemma 1.31), we can show that the Dunkl operator iT is symmetric.

Lemma 1.31.
The operator iT is symmetric and densely defined on $L^2(\mathbb{R}, \tilde{w})$.

Proof. The space $C_c^\infty(\mathbb{R})$ of compactly supported C^∞-functions is dense in $L^2(\mathbb{R}, \tilde{w})$. By Remark 1.30, $C_c^\infty(\mathbb{R})$ is a subset of $\mathcal{D}(T)$. Thus, also $\mathcal{D}(T)$ is a dense subset of $L^2(\mathbb{R}, \tilde{w})$. Finally, for $f, g \in \mathcal{D}(T)$, we just have to follow the lines of the proof of Lemma 1.16 to get the symmetry of iT. □

On the Hilbert space $L^2(\mathbb{R}, \tilde{w})$, we can now define the operators A and B as

$$Ag(t) = tg(t), \quad \mathcal{D}(A) = \left\{ g \in L^2(\mathbb{R}, \tilde{w}) : tg \in L^2(\mathbb{R}, \tilde{w}) \right\}, \tag{1.57}$$

$$Bg = iTg, \quad \mathcal{D}(B) = \mathcal{D}(T). \tag{1.58}$$

By Lemma 1.31, $B = iT$ is symmetric, and for $f, g \in \mathcal{D}(A)$, we get

$$\langle Af, g \rangle_{\tilde{w}} = \int_{\mathbb{R}} tf(t)\overline{g(t)}\tilde{w}(t)dt = \int_{\mathbb{R}} f(t)\overline{tg(t)}\tilde{w}(t)dt = \langle f, Ag \rangle_{\tilde{w}}.$$

Therefore, also A is symmetric. Even more, it is possible to show (cf. [85], Example (b), p. 318) that the operator A is self-adjoint. Now, adopting Theorem 1.2, we get the following uncertainty principle (cf. [27, (2.20)]):

Theorem 1.32.
Suppose that w satisfies Assumption 1.25. Let $f \in L^2([0,\infty), w) \cap AC_{loc}([0,\infty))$ such that $f', tf, tf', t\frac{w'}{w}f \in L^2([0,\infty), w)$. Further, let f be normalized such that $\|f\|_w = 1$. Then, the following uncertainty principle holds:

$$\|tf\|_w^2 \cdot \|f'\|_w^2 \geq \frac{1}{4}\left| 1 + \int_0^\infty tw'(t)|f(t)|^2 dt \right|^2. \tag{1.59}$$

Equality holds if and only if $f = Ce^{-\lambda t^2}$, with a complex scalar C and a real constant $\lambda \in \mathbb{R}$ such that f fulfills the requirements of the theorem.

Proof. We consider the operators A and B defined in (1.57) and (1.58) on the Hilbert space $L^2(\mathbb{R}, \tilde{w})$. The commutator of A and B is given by

$$[A, B]g(t) = -ig(t) - i\frac{\tilde{w}'(t)}{\tilde{w}(t)}tg(-t), \quad \text{for a.e. } t \in \mathbb{R},$$

for all functions $g \in \mathcal{D}(AB) \cap \mathcal{D}(BA)$. If we consider only even functions $g \in L^2_e(Y^d, \tilde{W})$, we get

$$[A, B]g(t) = -ig(t)\left(1 + t\frac{\tilde{w}'(t)}{\tilde{w}(t)}\right), \quad \text{for a.e. } t \in \mathbb{R}.$$

Next, if $f \in AC_{loc}([0, \infty))$ with f', tf, tf' and $t\frac{w'}{w}f$ in $L^2([0, \infty), w)$, then the even extension $e(f)$ lies in the domain $\mathcal{D}([A, B]) = \mathcal{D}(AB) \cap \mathcal{D}(BA)$ of the commutator. Hence, if we apply the symmetric operators A and B to the function $e(f)$, we get in inequality (1.7):

$$\|Ae(f)\|^2_{\tilde{w}} = \|te(f)\|^2_{\tilde{w}} = \|tf\|^2_w,$$
$$a = \langle Ae(f), e(f)\rangle_{\tilde{w}} = \langle te(f), e(f)\rangle_{\tilde{w}} = 0,$$
$$\|Be(f)\|_{\tilde{w}} = \|e(f)'\|_{\tilde{w}} = \|f'\|_w,$$
$$b = \langle Be(f), e(f)\rangle_{\tilde{w}} = \langle ie(f)', e(f)\rangle_{\tilde{w}} = 0,$$
$$[A, B]e(f)(t) = -ie(f)(t)\left(1 + t\frac{\tilde{w}'(t)}{\tilde{w}(t)}\right), \quad \text{for a.e. } t \in \mathbb{R}.$$

Further, since $\tilde{w}'$ is an odd function and f is normalized, we conclude

$$|\langle [A, B]e(f), e(f)\rangle_{\tilde{w}}| = \left|\int_{-\infty}^{\infty}\left(1 + \frac{\tilde{w}'(t)}{\tilde{w}(t)}t\right)|e(f)(t)|^2\tilde{w}(t)dt\right|$$
$$= \left|1 + \int_0^{\infty} tw'(t)|f(t)|^2dt\right|.$$

Due to Theorem 1.2, equality in (1.59) holds if and only if $ie(f)' = -i2\lambda te(f)$, where λ is a real constant. The solution of this differential equation is exactly the Gaussian function $e(f)(t) = Ce^{-\lambda t^2}$ with a complex scalar C. Restricted to the nonnegative real half axis, this yields the assertion. □

Remark 1.33. Theorem 1.24 in Section 1.4.2 can be considered as an intermediate result related to Theorem 1.18 in Section 1.4.1 and also to Theorem 1.32 above. Moreover, it is possible to prove Theorem 1.24 in two different ways. The first way, presented in Section 1.4.2, is by restricting the domain of the Dunkl operator T to functions with zero boundary condition. Alternatively, Theorem 1.24 can be proven by restricting the uncertainty inequality (1.59) above to functions that have compact support in the interval $[0, \pi]$.

1.5. Uncertainty principles for orthogonal expansions

If the weight function w determines the orthogonality measure of a set of orthogonal polynomials, the theory presented in the last section leads directly to an uncertainty relation for functions having an expansion in terms of orthogonal polynomials. The first work in this area was done by Rösler and Voit [73] who used the Dunkl operator T to develop an uncertainty principle for ultraspherical polynomials. Later on, these results were generalized by Li and Liu [54] to Jacobi polynomials. Using similar techniques as for the Jacobi polynomials, there have been developed uncertainty principles also for other well known orthogonal expansions like the spherical Bessel functions [74], the Laguerre polynomials and the generalized Hermite polynomials [55]. In this section, we will focus on the uncertainty principles for Jacobi and Laguerre expansions.

1.5.1. Uncertainty principles for Jacobi expansions

For $\alpha, \beta > -1$,the weight function $w_{\alpha\beta}$ of the Jacobi polynomials is defined on $[0, \pi]$ as

$$w_{\alpha\beta}(t) := 2^{\alpha+\beta+1} \sin^{2\alpha+1}\left(\frac{t}{2}\right) \cos^{2\beta+1}\left(\frac{t}{2}\right). \tag{1.60}$$

Integrating the weight function $w_{\alpha\beta}$ from 0 to π yields (cf. [42, (4.0.2)])

$$\int_0^\pi w_{\alpha\beta}(t)dt = 2^{\alpha+\beta+1}\frac{\Gamma(\alpha+1)\Gamma(\beta+1)}{\Gamma(\alpha+\beta+2)}. \tag{1.61}$$

If $\alpha, \beta \geq -\frac{1}{2}$, the Jacobi weight $w_{\alpha\beta}$ is a nonnegative and absolutely continuous function on $[0, \pi]$. Moreover, we have

$$\frac{w'_{\alpha\beta}(t)}{w_{\alpha\beta}(t)} = \frac{(\alpha+\beta+1)+(\alpha-\beta)\cos t}{\sin t}. \tag{1.62}$$

Hence, the Jacobi weight function $w_{\alpha\beta}$ satisfies Assumption 1.9.

The Jacobi polynomials $P_n^{(\alpha,\beta)}$ on the interval $[-1, 1]$ can be defined by the explicit formula (cf. [83, (4.21.1)])

$$P_n^{(\alpha,\beta)}(x) := \frac{\Gamma(n+\alpha+1)}{n!\Gamma(n+\alpha+\beta+1)} \sum_{j=0}^n \binom{n}{j} \frac{\Gamma(n+j+\alpha+\beta+1)}{\Gamma(j+\alpha+1)} \left(\frac{x-1}{2}\right)^j. \tag{1.63}$$

Using the coordinate change $[0, \pi] \to [-1, 1] : t \to \cos t = x$, we consider the polynomials $P_n^{(\alpha,\beta)}$ on the interval $[0, \pi]$. Then, the Jacobi polynomials $P_n^{(\alpha,\beta)}$ satisfy the following orthogonality relation (cf. [42, Theorem 4.1.1])

$$\int_0^\pi P_m^{(\alpha,\beta)}(\cos t) P_n^{(\alpha,\beta)}(\cos t) w_{\alpha\beta}(t)dt = h_n^{(\alpha,\beta)}\delta_{m,n}, \tag{1.64}$$

where

$$h_n^{(\alpha,\beta)} := \frac{1}{\|P_n^{(\alpha,\beta)}\|_{w_{\alpha\beta}}^2} = \frac{2^{\alpha+\beta+1}\Gamma(\alpha+n+1)\Gamma(\beta+n+1)}{n!\Gamma(\alpha+\beta+n+1)(\alpha+\beta+2n+1)}. \tag{1.65}$$

The Jacobi polynomials $\{P_n^{(\alpha,\beta)}(\cos t)\}_{n=0}^\infty$ form a complete orthogonal set in the Hilbert space $L^2([0,\pi], w_{\alpha\beta})$ (follows from the Stone-Weierstrass Theorem B.1 and the fact that the continuous functions on $[0,\pi]$ are dense in $L^2([0,\pi], w_{\alpha\beta})$). Moreover, they satisfy the second-order differential equation [83, (4.2.1)]

$$L_{\alpha\beta}P_n^{(\alpha,\beta)} = -n(n+\alpha+\beta+1)P_n^{(\alpha,\beta)}, \tag{1.66}$$

where the differential operator $L_{\alpha\beta}$ is given by

$$L_{\alpha\beta} := \frac{d^2}{dt^2} + \frac{(\alpha+\beta+1)\cos t + \alpha - \beta}{\sin t}\frac{d}{dt}. \tag{1.67}$$

By Definition 1.14, the Dunkl operator T on the symmetrically extended Hilbert space $L^2([-\pi,\pi], \tilde{w}_{\alpha\beta})$ is given by

$$Tg(t) = g'(t) + \frac{(\alpha+\beta+1)\cos t + (\alpha-\beta)}{\sin t}\frac{g(t)-g(-t)}{2}, \quad g \in \mathcal{D}(T). \tag{1.68}$$

The differential-difference operator T is strongly related to the second-order differential operator $L_{\alpha\beta}$. Namely, for functions $f \in C^2([0,\pi])$ with $f'(0) = f'(\pi) = 0$, we get

$$-L_{\alpha\beta}f = r((iT)^2 e(f)). \tag{1.69}$$

Therefore, the operator iT can be seen as a generalized root of the second-order differential operator $L_{\alpha\beta}$ and the frequency variance $\text{var}_F^{\alpha\beta}(f)$ of f can be written as

$$\text{var}_F^{\alpha\beta}(f) = \|f'\|_{w_{\alpha\beta}}^2 = \|iTe(f)\|_{\tilde{w}_{\alpha\beta}}^2 = \langle -L_{\alpha,\beta}f, f\rangle_{w_{\alpha\beta}}. \tag{1.70}$$

Finally, if we introduce the generalized mean value $\varepsilon_{\alpha\beta}(f)$ as in (1.38) by

$$\varepsilon_{\alpha\beta}(f) = \int_0^\pi \cos t\, |f(t)|^2 w_{\alpha\beta}(t)dt, \tag{1.71}$$

we can deduce from Corollary 1.42 an uncertainty principle for functions in $L^2([0,\pi], w_{\alpha\beta})$ (see also [54, Corollary 2] for an alternative formulation).

Corollary 1.34 (Uncertainty principle for Jacobi expansions).
Let $f \in AC([0,\pi]) \cap L^2([0,\pi], w_{\alpha\beta})$ with $f' \in L^2([0,\pi], w_{\alpha\beta})$, $\|f\|_{w_{\alpha\beta}} = 1$ and

$$(\alpha-\beta) + (\alpha+\beta+2)\varepsilon_{\alpha\beta}(f) \neq 0.$$

Then, the following uncertainty principle holds:

$$\frac{1-\varepsilon_{\alpha\beta}(f)^2}{|\frac{\alpha-\beta}{\alpha+\beta+2} + \varepsilon_{\alpha\beta}(f))|^2} \cdot \|f'\|_{w_{\alpha\beta}}^2 > \frac{(\alpha+\beta+2)^2}{4}. \tag{1.72}$$

Proof. Corollary 1.34 is an immediate consequence of Corollary 1.21 when applied to the weight function $w_{\alpha\beta}$. The only thing that remains to check is the formula for the integral term on the right hand side of equation (1.37). This is done by the following simple calculation:

$$\int_0^\pi \left(\cos t\, w_{\alpha\beta}(t) + \sin t\, w'_{\alpha\beta}(t) \right) |f(t)|^2 dt$$
$$= \int_0^\pi \left((\alpha + \beta + 2) \cos t + (\alpha - \beta) \right) |f(t)|^2 w_{\alpha\beta}(t) dt$$
$$= (\alpha - \beta) + (\alpha + \beta + 2)\varepsilon_{\alpha\beta}(f). \qquad \square$$

If $\alpha = \beta$, the polynomials $P_n^{(\alpha,\alpha)}$ are called ultraspherical polynomials. In this case, Corollary 1.34 corresponds to the original result of Rösler and Voit [73, Theorem 2.2].

Corollary 1.35 (Uncertainty principle for ultraspherical expansions).
Let $f \in L^2([0,\pi], w_{\alpha\alpha}) \cap AC([0,\pi])$ with $f' \in L^2([0,\pi], w_{\alpha\alpha})$, $\varepsilon_{\alpha\alpha}(f) \neq 0$ and normalized such that $\|f\|_{w_{\alpha\alpha}} = 1$. Then, the following uncertainty principle holds:

$$\frac{1 - \varepsilon_{\alpha\alpha}(f)^2}{\varepsilon_{\alpha\alpha}(f)^2} \cdot \|f'\|^2_{w_{\alpha\alpha}} > (\alpha + 1)^2. \qquad (1.73)$$

Remark 1.36. In the articles [54, Lemma 8] and [73, Proposition 3.3], it is shown that the constant $\frac{(\alpha+\beta+2)^2}{4}$ on the right hand side of inequality (1.72) is optimal. An alternative proof of this optimality will be given in Theorem 3.16 where we will show that for a family $\tilde{V}_n$ of polynomial kernels of order n the uncertainty product in (1.72) tends to the optimal constant as $n \to \infty$.

1.5.2. Uncertainty principles for Laguerre expansions

As an example for an orthogonal expansion on the nonnegative real half axis $[0, \infty)$, we consider the Laguerre polynomials. For $\alpha > -1$ the Laguerre polynomials L_n^α can be defined by the formula

$$L_n^{(\alpha)}(t) := \frac{e^t}{n!} \frac{d^n}{dt^n}(t^n e^{-t}) \qquad (1.74)$$

and satisfy the orthogonality relation [42, Theorem 4.6.1]

$$\int_0^\infty L_m^{(\alpha)}(t) L_n^{(\alpha)}(t) t^\alpha e^{-t} dt = \frac{\Gamma(\alpha + n + 1)}{n!} \delta_{m,n}. \qquad (1.75)$$

Hence, the Laguerre polynomials are orthogonal on $[0, \infty)$ with respect to the weight function $w_\alpha(t) = t^\alpha e^{-t}$. It is well-known that they form a complete orthogonal set in the Hilbert space $L^2([0, \infty), w_\alpha)$ and satisfy the second-order differential equation [42, (4.6.15)]

$$L_\alpha L_n^{(\alpha)} = -n L_n^{(\alpha)}, \qquad (1.76)$$

where the differential operator L_α is given by

$$L_\alpha := t\frac{d^2}{dt^2} + (1 + \alpha - t)\frac{d}{dt}. \tag{1.77}$$

For the fraction $\frac{w'_\alpha}{w_\alpha}$, we get in the case of the Laguerre polynomials

$$\frac{w'_\alpha(t)}{w_\alpha(t)} = \frac{\alpha}{t} - 1, \quad t \in (0, \infty).$$

Thus, if $\alpha \geq 0$, the weight function $w_{\alpha\beta}$ satisfies Assumption 1.25 and, by Definition 1.55, the Dunkl operator T on the extended Hilbert space $L^2(\mathbb{R}, \tilde{w}_\alpha)$ is given by

$$Tg(t) = g'(t) + \left(\frac{\alpha}{t} - \text{sign}\, t\right)\frac{g(t) - g(-t)}{2}, \quad \text{for a.e. } t \in \mathbb{R}, \quad g \in \mathcal{D}(T). \tag{1.78}$$

Also this time, the Dunkl operator T is related to the second order differential operator L_α. For functions $f \in C^2([0, \infty))$ with $f'(0) = 0$, we have

$$L_\alpha f = r\Big[(T \circ |t| \circ T)e(f)\Big]. \tag{1.79}$$

So, in the case of the Laguerre polynomials, the operator L_α can be decomposed with help of the Dunkl operator T and the multiplication operator $f \to |t|f$. Finally, as a consequence of Theorem 1.32, we get the following uncertainty principle for functions in $L^2([0, \infty), w_\alpha)$:

Corollary 1.37.
For $\alpha \geq 0$, let $f \in L^2([0, \infty), w_\alpha) \cap AC_{loc}([0, \infty))$ such that $f', tf, tf' \in L^2([0, \infty), w_\alpha)$. Further, let f be normalized such that $\|f\|_{w_\alpha} = 1$. Then, the following uncertainty inequality holds:

$$\|tf\|_{w_\alpha}^2 \cdot \|f'\|_{w_\alpha}^2 \geq \frac{1}{4}\left|1 + \alpha - \int_0^\infty t|f(t)|^2 w_\alpha(t)dt\right|^2. \tag{1.80}$$

Equality holds if and only if $f(t) = Ce^{-\lambda t^2}$ with a complex scalar C and a nonnegative real value $\lambda \geq 0$.

1.6. Remarks and References

Uncertainty principles in Hilbert spaces. Among many other standard references, Theorem 1.1 can be found in [19, Theorem 1.34] and [29, Lemma 2.2.2]. The Theorems 1.2 and 1.4 as well as their proofs are taken from [79].

The Heisenberg-Pauli-Weyl uncertainty principle. The Heisenberg-Paul-Weyl inequality (1.10) goes back to the pathbreaking work [33] of Heisenberg in 1927. In his work,

Heisenberg gives a detailed description and physical interpretation of (1.10). The precise mathematical formulation of (1.10), though, can firstly be found in the article of Kennard [45] and in the book of Weyl (see [86], p. 77) who credits the result to Pauli.

Among other standard references, the Heisenberg-Pauli-Weyl-inequality can be found in the form of inequality (1.10) in [20, Theorem 1.1], [29, Theorem 2.2.1] and [79, Theorem 6.1]. In the form of inequality (1.9), it can be found in [19, Corollary 1.35] or in [70, Theorem 2.2.3].

The Breitenberger uncertainty principle. The inequalities (1.15) and (1.18) go back to the primary work of Breitenberger [6] in 1983. In a more mathematical form, these two inequalities have been restated by Narcowich and Ward [63] in 1996 using the angular momentum operator from physics. In the version of (1.15) and (1.18), the Breitenberger uncertainty principle can be found in [68] and [78], [79]. The optimality of the constant $\frac{1}{4}$ on the right hand side of inequality (1.18) was firstly proven by Prestin and Quak in the article [68].

Section 1.4.1. The results of Section 1.4.1 constitute a generalization of the work of Rösler and Voit [73] who firstly used a Dunkl operator of the form (1.29) to prove an uncertainty principle for functions on $[0, \pi]$ having an expansion in terms of ultraspherical polynomials. The main results of the section are taken from the survey article [79] of Selig and from the work [26, 27] of Goh and Goodman. Related results can be also found in the article [55] of Li and Liu.

Assumption 1.9 is a slightly modified version of the assumptions on the weight function w given in [27, p. 23].

The definition of the Dunkl operator T in (1.29) corresponds to the definition of the Dunkl operator in [27, p. 23]. The definition (1.30) of the domain $\mathcal{D}(T)$ is slightly more general than in [27, p. 23].

The proof of the symmetry of the Dunkl operator T in Lemma 1.16 is analog to the proof of [73, Lemma 3.1]. The proof of the self-adjointness of T in Lemma 1.16 is very similar to the proof of the self-adjointness of the differential operator $\frac{d}{dt}$ on $L^2([-\pi, \pi])$ (see [70, Lemma 2.3.1]).

Theorem 1.18 and 1.19 are generalized versions of Theorem 9.1 and 9.2 in [79], respectively. Inequality (1.42) corresponds to inequality (2.12) in [27].

Section 1.4.2. Section 1.4.2 is a new result and can be considered as an intermediate result between Section 1.4.1 and Section 1.4.3. Namely, Theorem 1.24 can be proven in two ways. One way to prove it is by restricting the domain of the Dunkl operator T to functions with zero boundary condition as shown in Section 1.4.2. The other way to prove it is by restricting Theorem 1.32 in Section 1.4.3 to functions with compact support in the interval $[0, \pi]$.

Section 1.4.3. The results of Section 1.4.3 are mainly taken from [27]. Related results can also be found in [55].

Assumption 1.25 summarizes in a slightly modified way the assumptions on the weight function w given in [27, p. 24].

The definition of the differential-difference operator T in (1.55) corresponds to the definition of the Dunkl operator T in [27, p. 24]. The definition (1.56) of the domain $\mathcal{D}(T)$ is slightly more general than in [27, p. 24].
Inequality (1.42) is a reformulation of inequality (2.19) in [27]. For the weight function $w_\alpha(t) = t^\alpha$, $\alpha \geq 0$ on $[0, \infty)$, inequality (1.42) reduces to the particular case of uncertainty inequalities for Hankel transforms considered in [74].

Uncertainty principle for Jacobi polynomials. Corollary 1.34 can be found, in a slightly modified form, in the article [54, Corollary 2]. The original version of Corollary 1.35 can be found in [73, Theorem 2.2]. A qualitative uncertainty principle for Jacobi polynomials can be found in the dissertation [18] of Fischer.

Uncertainty principle for Laguerre polynomials. A slightly different version of Corollary 1.37 can be found in [55, Theorem 9]. Similar to the Laguerre case, an uncertainty principle for generalized Hermite polynomials was also proven in [55, Theorem 10].

2

Uncertainty principles on Riemannian manifolds

This chapter includes the main statements of this thesis. We will combine the theory of uncertainty principles in Chapter 1 with the geometric structure of a Riemannian manifold. First, we will extend the theory of Section 1.4 to weighted L^2-spaces on a multi-dimensional cylindrical domain Z_π^d. Then, this extended theory is used to develop an uncertainty principle for square-integrable functions on arbitrary compact Riemannian manifolds M. The main tools in this step are the exponential map $\exp_p$ on the tangential space T_pM and an isometric isomorphism from $L^2(M)$ onto a weighted L^2-space on the cylindrical domain Z_π^d. In a similar way, we will prove uncertainty principles for compact star-shaped domains $\Omega \subset M$ with Lipschitz continuous boundary and for manifolds which are diffeomorphic to the Euclidean space $\mathbb{R}^d$. Finally, we will show that the developed uncertainty principles are asymptotically sharp in the case that the underlying manifold is compact, and sharp if M is diffeomorphic to $\mathbb{R}^d$.

2.1. Weighted L^2-inequalities on a multi-dimensional cylindrical domain

2.1.1. Inequalities in the compact case

In this section, we are going to generalize the uncertainty principle of Theorem 1.19 to weighted L^2-spaces where the underlying domain is not the interval $[0, \pi]$ but the

d-dimensional cylinder

$$Z_\pi^d := [0, \pi] \times \mathbb{S}^{d-1} = \{(t, \xi) : t \in [0, \pi], \xi \in \mathbb{S}^{d-1}\} \subset \mathbb{R}^{d+1}, \qquad (2.1)$$

where $\mathbb{S}^{d-1} := \left\{\xi \in \mathbb{R}^d : |\xi|^2 = \xi_1^2 + \cdots + \xi_d^2 = 1\right\}$ denotes the $(d-1)$-dimensional unit sphere in $\mathbb{R}^d$. The cylinder Z_π^d is a differentiable submanifold of $\mathbb{R}^{d+1}$ with left hand boundary

$$\partial_L Z_\pi^d := \{(0, \xi) : \xi \in \mathbb{S}^{d-1}\} \qquad (2.2)$$

and right hand boundary

$$\partial_R Z_\pi^d := \{(\pi, \xi) : \xi \in \mathbb{S}^{d-1}\}. \qquad (2.3)$$

A Riemannian structure on Z_π^d is given by the restriction of the Euclidean metric in $\mathbb{R}^{d+1}$ to Z_π^d. Hence, Z_π^d is a compact Riemannian manifold with boundary ∂Z_L and ∂Z_R. A canonical measure on Z_π^d is given by the product measure $dt d\mu(\xi)$, where μ denotes the standard surface measure on $\mathbb{S}^{d-1}$.

Assumption 2.1. As admissible weight functions on Z_π^d, we consider positive functions W satisfying the properties

(i) $\quad W \in C(Z_\pi^d): \quad W(\cdot, \xi) \in AC([0, \pi]) \quad$ for μ-a.e. $\xi \in \mathbb{S}^{d-1}$,

$$W' = \frac{\partial W}{\partial t} \in L^1(Z_\pi^d), \qquad (2.4)$$

(ii) $\quad W(t, \xi) > 0, \quad (t, \xi) \in (0, \pi) \times \mathbb{S}^{d-1}, \qquad (2.5)$

(iii) $\quad t(\pi - t)\dfrac{W'}{W} \in L^\infty(Z_\pi^d). \qquad (2.6)$

Assumption 2.1 can be seen as an extension of Assumption 1.9 onto the d-dimensional cylinder Z_π^d. The symbol $C(Z_\pi^d)$ denotes the space of all continuous functions on the compact manifold Z_π^d. Then, the first condition (2.4) says that the weight function W is absolutely continuous with respect to the variable t for μ-a.e. fixed unit vector $\xi \in \mathbb{S}^{d-1}$, and that the Radon-Nikodym derivative W' of W with respect to the variable t is an integrable function on Z_π^d. Hereby, the notion "for μ-a.e. $\xi \in \mathbb{S}^{d-1}$" means that the statement holds for all unit vectors ξ on $\mathbb{S}^{d-1}$ except a subset of μ-measure zero, where μ is the standard Riemannian measure on $\mathbb{S}^{d-1}$. The second condition (2.5) ensures that the weight function W is strictly positive on the interior $Z_\pi^d \setminus \{\partial_L Z_\pi^d, \partial_R Z_\pi^d\}$ of the cylinder Z_π^d. Finally, the condition (2.6) guarantees that the fraction $t(\pi - t)\frac{W'}{W}$ is essentially bounded on Z_π^d. Especially the last property (2.6) will play an important role in the upcoming definition of the Dunkl operator.

Example 2.2. Consider the weight function $W_\alpha : Z_\pi^d \to \mathbb{R}$, $W_\alpha(t, \xi) := \sin^{2\alpha+1} t$, $\alpha \geq -\frac{1}{2}$. The conditions (2.4) and (2.5) are obviously satisfied. Moreover,

$$\left| t(\pi - t)\frac{W_\alpha'(t, \xi)}{W_\alpha(t, \xi)} \right| = |(2\alpha + 1)t(\pi - t)\cot t| \leq (2\alpha + 1)\pi.$$

Hence, also condition (2.6) is satisfied. In Section 2.6.1, we will see that for $\alpha = \frac{d-2}{2}$ the weight function W_α is related to the exponential map on the unit sphere $\mathbb{S}^d$.

Definition 2.3. For an admissible weight function W on Z_π^d we denote by $L^2(Z_\pi^d, W)$ the Hilbert space of weighted square integrable functions on Z_π^d with the inner product

$$\langle f, g \rangle_W := \int_{S^{d-1}} \int_0^\pi f(t, \xi) \overline{g(t, \xi)} W(t, \xi) dt d\mu(\xi). \tag{2.7}$$

The space $L^2(Z_\pi^d, W)$ is well defined and complete in the topology induced by the norm $\| \cdot \|_W^2 := \langle \cdot, \cdot \rangle_W$. This follows from the general theory of L^p-spaces on measure spaces (see Section B.1 of the appendix).

Similar as in Section 1.4, the differential operator $\frac{\partial}{\partial t}$ is in general not a symmetric operator on the Hilbert space $L^2(Z_\pi^d, W)$. This difficulty can be solved by introducing an appropriate differential-difference operator. To define such an operator, we have to double the domain of the Hilbert space $L^2(Z_\pi^d, W)$. This is done by doubling the range of the variable t.

Definition 2.4. On the doubled cylinder

$$X^d := [-\pi, \pi] \times S^{d-1} \subset \mathbb{R}^{d+1} \tag{2.8}$$

we define the extended weight function $\tilde{W}$ by

$$\tilde{W}(t, \xi) := \frac{1}{2} W(|t|, \xi), \quad (t, \xi) \in X^d. \tag{2.9}$$

By $L^2(X^d, \tilde{W})$, we denote the Hilbert space of weighted square integrable functions on X^d with the inner product

$$\langle f, g \rangle_{\tilde{W}} := \int_{S^d} \int_{-\pi}^\pi f(t, \xi) \overline{g(t, \xi)} \tilde{W}(t, \xi) dt d\mu(\xi). \tag{2.10}$$

As the cylinder Z_π^d, also the doubled cylinder X^d can be considered as a compact Riemannian manifold with boundary

$$\partial_L X^d := \{(-\pi, \xi) : \xi \in S^{d-1}\}, \tag{2.11}$$

$$\partial_R X^d := \{(\pi, \xi) : \xi \in S^{d-1}\}. \tag{2.12}$$

As the Hilbert space $L^2(Z_\pi^d, W)$, also the space $L^2(X^d, \tilde{W})$ is a well defined and complete Hilbert space with the norm $\| \cdot \|_{\tilde{W}}^2 := \langle \cdot, \cdot \rangle_{\tilde{W}}$ (see Section B.1).

Definition 2.5. On $L^2(X^d, \tilde{W})$, we define the reflection operator $\check{}$ by

$$\check{g}(t, \xi) := g(-t, \xi), \quad \text{for a.e. } (t, \xi) \in X^d.$$

We say that a function $g \in L^2(X^d, \tilde{W})$ is even (odd) in the variable t if it satisfies $\check{g} = g$ ($\check{g} = -g$, respectively). The subspace of even functions in $L^2(X^d, \tilde{W})$ is denoted by

$$L_e^2(X^d, \tilde{W}) := \left\{ g \in L^2(X^d, \tilde{W}) : \check{g} = g \right\}. \tag{2.13}$$

Similar as before, the notion 'for a.e. $(t, \xi) \in X^{d}$' means that the statement holds for all $(t, \xi) \in X^d$ except a set of measure zero, where in this case the measure is given by the canonical product measure $dt d\mu(\xi)$ on Z_π^d.

To relate the original Hilbert space $L^2(Z_\pi^d, W)$ with $L^2(X^d, \tilde{W})$, we introduce the even extension operator e and the restriction operator r as

$$e : L^2(Z_\pi^d, W) \to L^2(X^d, \tilde{W}), \quad e(f)(t, \xi) := f(|t|, \xi), \tag{2.14}$$

$$r : L^2(X^d, \tilde{W}) \to L^2(Z_\pi^d, W), \quad r(g) := g|_{[0,\pi] \times \mathbb{S}^{d-1}}. \tag{2.15}$$

Then, the operators e and r constitute isometric isomorphisms between the Hilbert space $L^2(Z_\pi^d, W)$ and the subspace $L_e^2(X^d, \tilde{W})$ of even functions.

By assumption (2.4), the weight function W is continuous on Z_π^d, $W(\cdot, \xi)$ is absolutely continuous for μ-a.e. $\xi \in \mathbb{S}^{d-1}$ and the Radon-Nikodym derivative W' is integrable on Z_π^d. Thus, also the symmetrically extended function $\tilde{W}$ is continuous on X^d and $\tilde{W}(\cdot, \xi) \in AC_{2\pi}$ is absolutely continuous for μ-a.e. $\xi \in \mathbb{S}^{d-1}$. Moreover, its Radon-Nikodym derivative with respect to the variable t, denoted as $\tilde{W}' := \frac{\partial \tilde{W}}{\partial t}$, is integrable on the doubled cylinder X^d.

Similar as in Definition 1.14, we can now introduce a differential-difference operator on $L^2(X^d, \tilde{W})$, referred to as Dunkl operator.

Definition 2.6. For a weight function W satisfying Assumption 2.15, we define the Dunkl operator T^X on the Hilbert space $L^2(X^d, \tilde{W})$ by

$$T^X g := \kappa \left(\frac{\partial g}{\partial t} + \frac{\tilde{W}'}{\tilde{W}} \frac{g - \check{g}}{2} \right), \tag{2.16}$$

where $\kappa \in C(\mathbb{S}^{d-1})$ denotes a strictly positive and continuous scaling function depending on the variable $\xi \in \mathbb{S}^{d-1}$. As a domain of T^X, we define

$$\mathcal{D}(T^X) := \Big\{ g \in L^2(X^d, \tilde{W}) : g(\cdot, \xi) \in AC_{2\pi} \quad \text{for } \mu\text{-a.e. } \xi \in \mathbb{S}^{d-1},$$

$$\frac{\partial g}{\partial t}, \frac{\tilde{W}'}{\tilde{W}} \frac{g - \check{g}}{2} \in L^2(X^d, \tilde{W}) \Big\}. \tag{2.17}$$

The function κ in (2.16) is an additional scaling function that allows to regulate the dependency on the variable $\xi \in \mathbb{S}^{d-1}$ in the Dunkl operator. In particular, such a scaling function κ will appear later on when we deal with uncertainty principles on compact Riemannian manifolds. The definition (2.17) of the domain $\mathcal{D}(T^X)$ is very similar to the definition (1.30) in the one-dimensional case. In principle, the domain $\mathcal{D}(T^X)$ consists of all functions g that are absolutely continuous for μ-a.e. fixed $\xi \in \mathbb{S}^{d-1}$ such that the Radon-Nikodym derivative $\frac{\partial g}{\partial t}$ and the fraction $\frac{\tilde{W}'}{\tilde{W}} \frac{g - \check{g}}{2}$ are elements of $L^2(X^d, \tilde{W})$.

Remark 2.7. Assumption 2.1 and, in particular, property (2.6) of the weight function W are crucial for Definition 2.6. These conditions ensure that the domain $\mathcal{D}(T^X)$ of the Dunkl operator is dense in $L^2(X^d, \tilde{W})$. To prove this, we consider the set

$$C^{1,t}_{2\pi}(X^d) := \left\{ g \in C(X^d) : \frac{\partial g}{\partial t} \in C(X^d),\ g(-\pi, \xi) = g(\pi, \xi),\ \frac{\partial g}{\partial t}(-\pi, \xi) = \frac{\partial g}{\partial t}(\pi, \xi) \right\} \tag{2.18}$$

and show that $C^{1,t}_{2\pi}(X^d)$ is a subset of $\mathcal{D}(T^X)$. Hereby, the t in the exponent indicates that we consider functions that are continuously differentiable with respect to the variable t. Obviously, if $g \in C^{1,t}_{2\pi}(X^d)$, then $g(\cdot, \xi)$ is absolutely continuous for all $\xi \in \mathbb{S}^{d-1}$ and, since X^d is compact, $\frac{\partial g}{\partial t} \in L^2(X^d, \tilde{W})$. It remains to check that $\frac{\tilde{W}'}{\tilde{W}} \frac{g - \check{g}}{2} \in L^2(X^d, \tilde{W})$. Since $\frac{\partial g}{\partial t} \in C(X^d)$ and $g(-\pi, \xi) = g(\pi, \xi)$ for all $\xi \in \mathbb{S}^{d-1}$, the function $\frac{g - \check{g}}{2t(\pi^2 - t^2)}$ is continuous on X^d. Thus, we get

$$\int_{\mathbb{S}^{d-1}} \int_{-\pi}^{\pi} \left| \frac{\tilde{W}'(t, \xi)}{\tilde{W}(t, \xi)} \frac{g(t, \xi) - g(-t, \xi)}{2} \right|^2 \tilde{W}(t, \xi)\, dt\, d\mu(\xi)$$

$$= \int_{\mathbb{S}^{d-1}} \int_{-\pi}^{\pi} \left| t(\pi^2 - t^2) \frac{\tilde{W}'(t, \xi)}{\tilde{W}(t, \xi)} \right|^2 \left| \frac{g(t, \xi) - g(-t, \xi)}{2t(\pi^2 - t^2)} \right|^2 \tilde{W}(t, \xi)\, dt\, d\mu(\xi)$$

$$\leq \left\| t(\pi^2 - t^2) \frac{W'}{W} \right\|^2_{L^\infty(Z^d_\pi)} \left\| \frac{g - \check{g}}{2t(\pi^2 - t^2)} \right\|^2_{\tilde{W}}.$$

Since the weight function W satisfies property (2.6), we conclude that the fraction $\frac{\tilde{W}'}{\tilde{W}} \frac{g - \check{g}}{2}$ is in $L^2(X^d, \tilde{W})$.

Similar as in Lemma 1.16 for the one-dimensional setting, we can now show that the Dunkl operator T^X is symmetric on $L^2(X^d, \tilde{W})$.

Lemma 2.8.
The operator iT^X with domain $\mathcal{D}(T^X)$ is symmetric and densely defined on $L^2(X^d, \tilde{W})$.

Proof. The set

$$C_{2\pi}(X^d) := \{ g \in C(X^d) : g(-\pi, \xi) = g(\pi, \xi) \}$$

is dense in $L^2(X^d, \tilde{W})$ (follows, for instance, from [38, Theorem 13.21]). Hence, for every $g \in L^2(X^d, \tilde{W})$ and $\epsilon > 0$ there exists a $g_c \in C_{2\pi}(X^d)$ such that $\|g - g_c\|_{\tilde{W}} < \epsilon$. Further, the space $C^{1,t}_{2\pi}(X^d)$ defined in (2.18) is a subalgebra of $C_{2\pi}(X^d)$ that contains the constant functions, separates the points on $X^d \setminus \partial_L X^d$ and is closed under complex conjugation. To see that $C^{1,t}_{2\pi}(X^d)$ separates the points on $X^d \setminus \partial_L X^d$, we define for $p = (t_1, \xi_1), q = (t_2, \xi_2) \in X^d \setminus \partial_L X^d$, $t_1 - t_2 \neq \{0, \pm\pi\}$, the function $s_1(t, \xi) = \frac{\sin(t - t_1)}{\sin(t_2 - t_1)}$. Then, $s_1 \in C^{1,t}_{2\pi}(X^d)$ and $s_1(p) = 0$, $s_1(q) = 1$. If $t_1 - t_2 = \pm\pi$, we define $s_2(t, \xi) = \frac{\cos(t - t_1) + 1}{2}$. Then, also $s_2 \in C^{1,t}_{2\pi}(X^d)$ and $s_2(p) = 0$, $s_2(q) = 1$. If $t_1 = t_2$, $\xi_1 \neq \xi_2$, we take as separating function $s_3(t,\xi) = r(\xi)$, where r is a continuous function on $\mathbb{S}^{d-1}$ with $r(\xi_1) = 0$ and $r(\xi_2) = 1$.

Hence, by the Stone-Weierstrass Theorem B.1, $C_{2\pi}^{1,t}(X^d)$ is dense in $C_{2\pi}(X^d)$ and for every $g_c \in C_{2\pi}(X^d)$ there exists a $g_1 \in C_{2\pi}^{1,t}(X^d)$ such that $\|g_c - g_1\|_\infty < \epsilon$. Moreover, by Remark 2.7, the function g_1 is an element of $\mathcal{D}(T^X)$. In total, we get

$$\|g - g_1\|_{\tilde{W}} \leq \|g - g_c\|_{\tilde{W}} + \|g_c - g_1\|_{\tilde{W}}$$
$$< \epsilon + \|g_c - g_1\|_\infty \int_{\mathbb{S}^{d-1}} \int_{-\pi}^{\pi} \tilde{W}(t,\xi) dt d\mu(\xi) < (1 + \|W\|_{L^1(Z_\pi^d)})\epsilon.$$

Thus, $\mathcal{D}(T^X)$ is a dense subset of $L^2(X^d, \tilde{W})$.

To check the symmetry of iT^X, we essentially follow the lines of the proof of Lemma 1.16. For $f, g \in \mathcal{D}(T^X)$, using integration by parts with respect to the variable t, we get the identity

$$\int_{\mathbb{S}^{d-1}} \int_{-\pi}^{\pi} \frac{\partial f}{\partial t}(t,\xi)\overline{g(t,\xi)}\tilde{W}(t,\xi) dt d\mu(\xi) = -\int_{\mathbb{S}^{d-1}} \int_{-\pi}^{\pi} f(t,\xi)\frac{\partial}{\partial t}\Big(\overline{g(t,\xi)}\tilde{W}(t,\xi)\Big) dt d\mu(\xi)$$
$$= -\int_{\mathbb{S}^{d-1}} \int_{-\pi}^{\pi} f(t,\xi)\Big(\overline{\frac{\partial g}{\partial t}(t,\xi)} + \overline{g(t,\xi)}\frac{\tilde{W}'(t,\xi)}{\tilde{W}(t,\xi)}\Big)\tilde{W}(t,\xi) dt d\mu(\xi).$$

Now, by definition (2.16) of the operator T^X, we get

$$\int_{\mathbb{S}^{d-1}} \int_{-\pi}^{\pi} (iT^X f)(t,\xi)\overline{g(t,\xi)}\tilde{W}(t,\xi) dt d\mu(\xi)$$
$$= -i\int_{\mathbb{S}^{d-1}} \int_{-\pi}^{\pi} \kappa(\xi)\Big(f(t,\xi)\overline{\frac{\partial g}{\partial t}(t,\xi)} + f(t,\xi)\overline{g(t,\xi)}\frac{\tilde{W}'(t,\xi)}{\tilde{W}(t,\xi)}\Big)\tilde{W}(t,\xi) dt d\mu(\xi)$$
$$+ i\int_{\mathbb{S}^{d-1}} \int_{-\pi}^{\pi} \kappa(\xi)\frac{f(t,\xi) - f(-t,\xi)}{2}\overline{g(t,\xi)}\frac{\tilde{W}'(t,\xi)}{\tilde{W}(t,\xi)}\tilde{W}(t,\xi) dt d\mu(\xi)$$
$$= -i\int_{\mathbb{S}^{d-1}} \int_{-\pi}^{\pi} \kappa(\xi)\Big(f(t,\xi)\overline{\frac{\partial g}{\partial t}(t,\xi)} + \frac{f(t,\xi) + f(-t,\xi)}{2}\overline{g(t,\xi)}\frac{\tilde{W}'(t,\xi)}{\tilde{W}(t,\xi)}\Big)\tilde{W}(t,\xi) dt d\mu(\xi)$$
$$= -i\int_{\mathbb{S}^{d-1}} \int_{-\pi}^{\pi} \kappa(\xi)\Big(f(t,\xi)\overline{\frac{\partial g}{\partial t}(t,\xi)} + f(t,\xi)\frac{\overline{g(t,\xi)} - \overline{g(-t,\xi)}}{2}\frac{\tilde{W}'(t,\xi)}{\tilde{W}(t,\xi)}\Big)\tilde{W}(t,\xi) dt d\mu(\xi)$$
$$= \int_{\mathbb{S}^{d-1}} \int_{-\pi}^{\pi} f(t,\xi)\overline{(iT^X g)(t,\xi)}\tilde{W}(t,\xi) dt d\mu(\xi).$$

Hence, the operator T^X is symmetric on the domain $\mathcal{D}(T^X)$. $\qquad\square$

Now, we fix a multiplier $h \in \mathcal{D}(T^X)$ and define two operators A and B on $L^2(X^d, \tilde{W})$ by

$$Ag := hg, \qquad \mathcal{D}(A) = L^2(X^d, \tilde{W}), \qquad\qquad (2.19)$$
$$Bg := iT^X g, \qquad \mathcal{D}(B) = \mathcal{D}(T^X). \qquad\qquad (2.20)$$

The multiplication operator A is a normal and bounded operator on $L^2(X^d, \tilde{W})$. The differential-difference operator B is symmetric due to Lemma 1.16. For these two operators, Theorem 1.4 implies

Theorem 2.9.
For a multiplier $h \in \mathcal{D}(T^X)$ and an even function $g \in L_e^2(X^d, \tilde{W}) \cap \mathcal{D}(T^X)$, the following inequality holds:

$$\left(\|hg\|_{\tilde{W}}^2 - \frac{|\langle hg, g \rangle_{\tilde{W}}|^2}{\|g\|_{\tilde{W}}} \right) \left\| \kappa \frac{\partial g}{\partial t} \right\|_{\tilde{W}}^2 \geq \frac{1}{4} |\langle g\, T^X h, g \rangle_{\tilde{W}}|^2. \qquad (2.21)$$

Proof. For an even function $g \in L_e^2(X^d, \tilde{W}) \cap \mathcal{D}(T^X)$, we have

$$T^X g(t, \xi) = \kappa(\xi) \frac{\partial g}{\partial t}(t, \xi) \quad \text{for a.e. } (t, \xi) \in X^d.$$

In particular, the function $T^X g$ satisfies $T^X g(t, \xi) = -T^X g(-t, \xi)$ for a.e. $(t, \xi) \in X^d$, implying that $\langle Bg, g \rangle_{\tilde{W}} = \langle iT^X g, g \rangle_{\tilde{W}} = 0$.
The commutator $[A, B]$ of A and B acting on functions $g \in \mathcal{D}(AB) \cap \mathcal{D}(BA) = \mathcal{D}(T^X)$ is given by

$$[A, B]g(t, \xi) = ihT^X g(t, \xi) - iT^X(hg)(t, \xi)$$
$$= -i\kappa(\xi) \left(\frac{\partial h}{\partial t}(t, \xi) g(t, \xi) + \frac{w'(t, \xi)}{w(t, \xi)} \frac{h(t, \xi) - h(-t, \xi)}{2} \right) g(-t, \xi).$$

Since $g \in L_e^2(X^d, \tilde{W}) \cap \mathcal{D}(T^X)$ is even, we get

$$[A, B]g = -(iT^X h)g. \qquad (2.22)$$

Now, inserting the values $a = \langle hg, g \rangle_{\tilde{W}}$, $b = \langle iT^X g, g \rangle_{\tilde{W}} = 0$ and identity (2.22) for the commutator $[A, B]$ in inequality (1.8), we get inequality (2.21). $\qquad \square$

An inequality for the initial Hilbert space $L^2(Z_\pi^d, W)$ can now be formulated by extending functions symmetrically onto the Hilbert space $L^2(X^d, \tilde{W})$ and using Theorem 2.9. In particular, for the subset

$$\mathcal{D}(\tfrac{\partial}{\partial t}; Z_\pi^d) := \Big\{ f \in L^2(Z_\pi^d, W) : f(\cdot, \xi) \in AC([0, \pi]) \text{ for } \mu\text{-a.e. } \xi \in \mathbb{S}^{d-1},$$
$$\frac{\partial f}{\partial t} \in L^2(Z_\pi^d, W) \Big\} \subset L^2(Z_\pi^d, W), \qquad (2.23)$$

we get

Theorem 2.10.
Suppose that $\kappa \in C(\mathbb{S}^{d-1})$ is a strictly positive scaling function on $\mathbb{S}^{d-1}$ and that the weight function W satisfies Assumption 2.1. Let $f \in L^2(Z_\pi^d, W) \cap \mathcal{D}(\tfrac{\partial}{\partial t}; Z_\pi^d)$ be normalized such that $\|f\|_W = 1$. Then, the following inequality holds:

$$\left(1 - \left(\int_{\mathbb{S}^{d-1}} \int_0^\pi \cos t \, |f(t, \xi)|^2 W(t, \xi) dt d\mu(\xi) \right)^2 \right) \cdot \left\| \kappa \frac{\partial f}{\partial t} \right\|_W^2 \geq \qquad (2.24)$$
$$\frac{1}{4} \left| \int_{\mathbb{S}^{d-1}} \int_0^\pi \kappa(\xi) \Big(\cos t \, W(t, \xi) + \sin t \, W'(t, \xi) \Big) |f(t, \xi)|^2 dt d\mu(\xi) \right|^2.$$

Equality in (2.24) is attained if and only if $f(\cdot, \xi) = C_\xi$ is constant for μ-a.e. $\xi \in \mathbb{S}^{d-1}$.

Proof. If $f \in \mathcal{D}(\frac{\partial}{\partial t}; Z_\pi^d)$, then $e(f)(\cdot, \xi) \in AC_{2\pi}$ for μ-a.e. $\xi \in \mathbb{S}^{d-1}$, $\|\frac{\partial e(f)}{\partial t}\|_{\tilde{W}} = \|\frac{\partial f}{\partial t}\|_W$ and $e(f) = e(f)^{\vee}$. Hence, the even extension $e(f) \in \mathcal{D}(T^X)$ is an element of the domain of the Dunkl operator T^X. Now, we adopt inequality (2.21) to prove (2.24). As a multiplier function h in the definition (2.19) of the operator A, we choose

$$h(t, \xi) = e^{it}, \quad (t, \xi) \in X^d.$$

Then, $h \in \mathcal{D}(T^X)$, and we have

$$\|he(f)\|_{\tilde{W}}^2 = \|e(f)\|_{\tilde{W}}^2 = \|f\|_W^2 = 1,$$

$$|\langle he(f), e(f) \rangle_{\tilde{W}}|^2 = |\langle e^{it}e(f), e(f) \rangle_{\tilde{W}}|^2 = \left(\int_{\mathbb{S}^{d-1}} \int_0^\pi \cos t |f(t, \xi)|^2 W(t, \xi) dt d\mu(\xi) \right)^2,$$

$$\|T^X e(f)\|_{\tilde{W}} = \left\| \kappa \frac{\partial e(f)}{\partial t} \right\|_{\tilde{W}} = \left\| \kappa \frac{\partial f}{\partial t} \right\|_W,$$

$$iT^X h(t, \xi) = \kappa(\xi) \left(-e^{it} - \frac{\tilde{W}'(t, \xi)}{\tilde{W}(t, \xi)} \sin t \right) \quad \text{for a.e. } (t, \xi) \in X^d.$$

Further, since $\sin t$ and $\frac{\tilde{W}'}{\tilde{W}}$ are odd functions in the variable t, we conclude

$$\langle e(f) i T^X h, e(f) \rangle_{\tilde{W}} = \int_{\mathbb{S}^{d-1}} \int_{-\pi}^\pi \kappa(\xi) \left(-e^{it} - \frac{\tilde{W}'(t, \xi)}{\tilde{W}(t, \xi)} \sin t \right) |e(f)(t, \xi)|^2 \tilde{W}(t, \xi) dt d\mu(\xi)$$

$$= -\int_{\mathbb{S}^{d-1}} \int_0^\pi \kappa(\xi) \left(\cos t W(t, \xi) + \sin t W'(t, \xi) \right) |f(t, \xi)|^2 dt d\mu(\xi).$$

Hence, we have shown that inequality (2.24) holds.

Due to Theorem 1.8, equality in (2.24) is attained if and only if for $a = \langle e^{it}e(f), e(f) \rangle_{\tilde{W}}$ and $b = \langle iT^X e(f), e(f) \rangle_{\tilde{W}} = 0$ the following identity holds:

$$i\kappa \frac{\partial e(f)}{\partial t} = \lambda(e^{it} - a)e(f) = -\bar{\lambda}(e^{-it} - \bar{a})e(f), \quad \lambda \in \mathbb{C}.$$

The second identity implies

$$e(f)(t, \xi) \left(\lambda e^{it} + \bar{\lambda} e^{-it} - a\lambda - \bar{\lambda}\bar{a} \right) = 2e(f)(t, \xi) \left(\text{Re}(\lambda e^{-it}) - \text{Re}(a\lambda) \right) = 0$$

for μ-a.e. $\xi \in \mathbb{S}^{d-1}$. This condition can only be satisfied if $e(f) = 0$ or if $\lambda = 0$. In the latter case we get $i\kappa \frac{\partial e(f)}{\partial t} = 0$ for μ-a.e. $\xi \in \mathbb{S}^{d-1}$. Hence, $e(f)(t, \xi) = C_\xi$ for μ-a.e. $\xi \in \mathbb{S}^{d-1}$ and, in particular, the function f does not depend on the variable t. $\square$

Inequality (2.24) can evidently be seen as a multi-dimensional version of the uncertainty principle (1.37) originally shown by Goh and Goodman in [27]. In both cases, the theory and the techniques are conceptually the same. The difference lies in the fact that in the higher-dimensional case above the weight function W and the Dunkl operator T^X additionally depend on a variable $\xi \in \mathbb{S}^{d-1}$.

Example 2.11. For the weight function $W_\alpha(t,\xi) = \sin^{2\alpha+1} t$, $\alpha \geq -1/2$, and the scaling function $\kappa = 1$ the Dunkl operator T^X on $L^2(X^d, \tilde{W}_\alpha)$ reads as

$$T^X g(t,\xi) = \frac{\partial g}{\partial t}(t,\xi) + (2\alpha + 1)\frac{\cos t}{\sin t}\frac{g(t,\xi) - g(-t,\xi)}{2}, \quad \text{for a.e. } (t,\xi) \in X^d.$$

In this case, inequality (2.24) attains the form

$$\left(1 - \left(\int_{\mathbb{S}^{d-1}}\int_0^\pi \cos t |f(t,\xi)|^2 W_\alpha(t,\xi) dt d\mu(\xi)\right)^2\right) \cdot \left\|\frac{\partial f}{\partial t}\right\|_W^2$$

$$\geq (1+\alpha)^2 \left(\int_{\mathbb{S}^{d-1}}\int_0^\pi \cos t |f(t,\xi)|^2 W_\alpha(t,\xi) dt d\mu(\xi)\right)^2.$$

This Dunkl operator is a multi-dimensional version of the Dunkl operator introduced by Rösler and Voit in [73] for ultraspherical expansions.

2.1.2. Inequalities in the compact case with zero boundary condition

The commutator $[A, B]$ in (2.22) is well defined for functions in $\mathcal{D}(T^X)$ if the multiplier function h is an element of $\mathcal{D}(T^X)$. This is evidently the case if h is given by $h(t,\xi) = e^{it}$, but not if we choose, for instance, $h(t,\xi) = t$. In the second case we have to restrict, similar as in Theorem 1.24, the domain of the Dunkl operator.

Definition 2.12. Let $L^2(X^d, \tilde{W})$ be the extension of $L^2(Z^d_\pi, W)$ as in Definition 2.4. We restrict the Dunkl operator T^X defined in (2.16) by $T^X g = \kappa(\frac{\partial g}{\partial t} + \frac{\tilde{W}'}{\tilde{W}}\frac{g-\check{g}}{2})$ to the smaller domain

$$\mathcal{D}_0(T^X) := \Big\{g \in L^2(X^d, \tilde{W}) : g(\cdot,\xi) \in AC_{2\pi},\ g(\pi,\xi) = 0 \text{ for } \mu\text{-a.e. } \xi \in \mathbb{S}^{d-1},$$

$$\frac{\partial g}{\partial t},\ \frac{\tilde{W}'}{\tilde{W}}\frac{g - \check{g}}{2},\ t\frac{\tilde{W}'}{\tilde{W}}g \in L^2(X^d, \tilde{W})\Big\}. \tag{2.25}$$

Since $\mathcal{D}_0(T^X) \subset \mathcal{D}(T^X)$, Lemma 2.8 implies that the operator iT^X is also symmetric on the smaller domain $\mathcal{D}_0(T^X)$. Put in another way, the operator iT^X on $\mathcal{D}(T^X)$ is a symmetric extension of $iT^X|_{\mathcal{D}_0(T^X)}$.

On $L^2(X^d, \tilde{W})$, we consider now the operator A defined by $Ag = hg$ and the operator $B = iT^X$ defined on the restricted domain $\mathcal{D}_0(T^X)$. In the following, we will show that the commutator $[A, B]$ is well defined for functions in $\mathcal{D}_0(T^X)$ if the multiplier function h is an element of

$$\mathcal{M}^X := \Big\{h \in L^2(X^d, \tilde{W}) : h(\cdot,\xi) \in AC([-\pi,\pi]) \text{ for } \mu\text{-a.e. } \xi \in \mathbb{S}^{d-1},$$

$$\frac{\partial h}{\partial t} \in L^2(X^d, \tilde{W}),\ (\pi^2 - t^2)\frac{\tilde{W}'}{\tilde{W}}\frac{h - \check{h}}{2}, \in L^2(X^d, \tilde{W})\Big\}. \tag{2.26}$$

We remark that the multipliers $h \in \mathcal{M}^X$, in contrast to functions $g \in \mathcal{D}(T^X)$ do not have to fulfill the periodicity condition $h(-\pi, \xi) = h(\pi, \xi)$ for μ-a.e. $\xi \in \mathbb{S}^{d-1}$ and that the condition $(\pi^2 - t^2) \frac{\tilde{W}'}{\tilde{W}} \frac{h - \check{h}}{2} \in L^2(X^d, \tilde{W})$ is weaker than the condition $\frac{\tilde{W}'}{\tilde{W}} \frac{g - \check{g}}{2} \in L^2(X^d, \tilde{W})$ in the domain $\mathcal{D}(T)$. Similar to Theorem 2.9, we get the following result for even functions in the restricted domain $\mathcal{D}_0(T^X)$.

Theorem 2.13.
For an even function $g \in L^2_e(X^d, \tilde{W}) \cap \mathcal{D}_0(T^X)$ *and a fixed multiplier* $h \in \mathcal{M}^X$ *the following inequality holds:*

$$\left(\|hg\|_{\tilde{W}}^2 - \frac{|\langle hg, g \rangle_{\tilde{W}}|^2}{\|g\|_{\tilde{W}}} \right) \cdot \left\| \kappa \frac{\partial g}{\partial t} \right\|_{\tilde{W}}^2 \geq \frac{1}{4} |\langle h(T^X g) - T^X (hg), g \rangle_{\tilde{W}}|^2. \tag{2.27}$$

Proof. The operator A defined by $Ag = hg$ is a normal and bounded operator on $L^2(X^d, \tilde{W})$ and the operator $B = iT^X$ defined on $\mathcal{D}(B) = \mathcal{D}_0(T^X)$ is symmetric due to Lemma 2.8. For an even $g \in L^2_e(X^d, \tilde{W}) \cap \mathcal{D}(T^X)$, we have $T^X g(t, \xi) = \kappa(\xi) \frac{\partial g}{\partial t}(t, \xi)$ for a.e. $(t, \xi) \in X^d$. Further, the fact that $\frac{\partial g}{\partial t}(t, \xi) = -\frac{\partial g}{\partial t}(-t, \xi)$ holds for a.e. $(t, \xi) \in X^d$ implies that $\langle Bg, g \rangle_{\tilde{W}} = 0$.
For the product hg of $g \in \mathcal{D}_0(T^X)$ and $h \in \mathcal{M}^X$, we have $hg(\cdot, \xi) \in AC_{2\pi}$, $hg(-\pi, \xi) = hg(\pi, \xi) = 0$ for μ-a.e. $\xi \in \mathbb{S}^{d-1}$ and $\frac{\partial hg}{\partial t} \in L^2(X^d, \tilde{W})$. Further,

$$\left\| \frac{\tilde{W}'}{\tilde{W}} \frac{hg - \check{h}\check{g}}{2} \right\|_{\tilde{W}}^2 = \left\| \frac{\tilde{W}'}{\tilde{W}} \frac{h - \check{h}}{2} \frac{g + \check{g}}{2} + \frac{\tilde{W}'}{\tilde{W}} \frac{g - \check{g}}{2} \frac{h + \check{h}}{2} \right\|_{\tilde{W}}^2$$

$$\leq \left\| (\pi^2 - t^2 + |t|) \frac{\tilde{W}'}{\tilde{W}} \frac{h - \check{h}}{2} \frac{g + \check{g}}{2} \right\|_{\tilde{W}}^2 + \left\| \frac{\tilde{W}'}{\tilde{W}} \frac{g - \check{g}}{2} \right\|_{\tilde{W}}^2 \|h\|_\infty$$

$$\leq \left\| (\pi^2 - t^2) \frac{\tilde{W}'}{\tilde{W}} \frac{h - \check{h}}{2} \right\|_{\tilde{W}}^2 \|g\|_\infty + \left\| t \frac{\tilde{W}'}{\tilde{W}} g \right\|_{\tilde{W}}^2 \|h\|_\infty + \left\| \frac{\tilde{W}'}{\tilde{W}} \frac{g - \check{g}}{2} \right\|_{\tilde{W}}^2 \|h\|_\infty.$$

Thus, $hg \in \mathcal{D}_0(T^X)$ and the commutator $[A, B]g = hiT^X g - iT^X(hg)$ is well defined for functions $g \in \mathcal{D}_0(T^X)$. Inequality (2.27) now follows from inequality (1.8) with $a = \langle hg, g \rangle_{\tilde{W}}$ and $b = \langle iT^X g, g \rangle_{\tilde{W}} = 0$. $\qquad\square$

Since $\mathcal{M}^X \supset \mathcal{D}(T^X)$, we are more flexible in the choice of the multiplier h in Theorem 2.13 than in Theorem 2.9. In particular, functions h of the type $h(t, \xi) = \iota(\xi)t$, where ι is a nonnegative and continuous scaling function on $\mathbb{S}^{d-1}$ are admissible multipliers in Theorem 2.13. Now, for functions f in

$$\mathcal{D}_0(\tfrac{\partial}{\partial t}; Z^d_\pi) := \Big\{ f \in L^2(Z^d_\pi, W) : f(\cdot, \xi) \in AC([0, \pi]), \ f(\pi, \xi) = 0 \quad \text{for } \mu\text{-a.e. } \xi \in \mathbb{S}^{d-1},$$

$$\frac{\partial f}{\partial t}, t \frac{W'}{W} f \in L^2(Z^d_\pi, w) \Big\} \subset L^2(Z^d_\pi, W), \tag{2.28}$$

we can derive the following inequality:

Theorem 2.14.

Assume that $\iota, \kappa \in C(\mathbb{S}^{d-1})$ are strictly positive scaling functions on $\mathbb{S}^{d-1}$, and that the weight function W satisfies Assumption 2.1. Let $f \in L^2(Z_\pi^d, W) \cap \mathcal{D}_0(\frac{\partial}{\partial t}; Z_\pi^d)$ be normalized such that $\|f\|_W = 1$. Then, the following inequality holds:

$$\|\iota\, t f\|_W^2 \cdot \left\|\kappa \frac{\partial f}{\partial t}\right\|_W^2 > \frac{1}{4}\left|\int_{\mathbb{S}^{d-1}} \int_0^\pi \iota(\xi)\kappa(\xi)(W(t,\xi) + tW'(t,\xi))|f(t,\xi)|^2 dt d\mu(\xi)\right|^2 \quad (2.29)$$

There is no function $f \in \mathcal{D}_0(\frac{\partial}{\partial t}; Z_\pi^d)$, $\|f\|_W = 1$, for which equality is attained in (2.29).

Proof. We proceed as in the proof of Theorem 2.10. If $f(\cdot, \xi) \in AC([0, \pi])$, $f(\pi, \xi) = 0$, for μ-a.e. $\xi \in \mathbb{S}^{d-1}$ and $\frac{\partial f}{\partial t}, t\frac{W'}{W}f \in L^2(Z_\pi^d, W)$, then the even extension $e(f) \in \mathcal{D}_0(T^X)$ and we can adopt Theorem 2.13. If we choose the multiplier function $h \in \mathcal{M}^X$ as $h(t,\xi) = \iota(\xi)t$, $(t,\xi) \in X^d$, then we get in inequality (2.27):

$$\|he(f)\|_{\tilde{W}}^2 = \|\iota\, te(f)\|_{\tilde{W}}^2 = \|\iota\, tf\|_W^2,$$

$$\langle he(f), e(f)\rangle_{\tilde{W}} = \langle \iota\, te(f), e(f)\rangle_{\tilde{W}} = 0,$$

$$\left\|\kappa \frac{\partial e(f)}{\partial t}\right\|_{\tilde{W}} = \left\|\kappa \frac{\partial f}{\partial t}\right\|_W,$$

$$\langle iT^X e(f), e(f)\rangle_{\tilde{W}} = \left\langle \kappa \frac{\partial e(f)}{\partial t}, e(f)\right\rangle_{\tilde{W}} = 0,$$

$$T^X(he(f)) - hT^X(e(f))) = \iota\kappa\left(1 + \frac{\tilde{W}'}{\tilde{W}}t\right)e(f).$$

Further, since t and $\frac{\tilde{W}'}{\tilde{W}}$ are odd functions in the variable t, we conclude

$$\langle -i[A, B]e(f), e(f)\rangle_{\tilde{W}} = \left\langle \iota\kappa\left(1 + \frac{\tilde{W}'}{\tilde{W}}t\right)e(f), e(f)\right\rangle_{\tilde{W}}$$

$$= \int_{\mathbb{S}^{d-1}} \int_0^\pi \iota(\xi)\kappa(\xi)\left(W(t,\xi) + tW'(t,\xi)\right)|f(t,\xi)|^2 dt d\mu(\xi).$$

Hence, inequality (2.29) is shown.

Since both operators A and B are symmetric, Theorem 1.2 states that equality in (2.29) is attained if and only if

$$i\kappa(\xi)\frac{\partial e(f)}{\partial t}(t,\xi) = i\lambda\iota(\xi)t\, e(f)(t,\xi), \quad \lambda \in \mathbb{R},$$

holds for a.e. $(t,\xi) \in X^d$, i.e., if and only if $f(t,\xi) = Ce^{-\lambda \frac{\iota(\xi)}{\kappa(\xi)}t^2}$ for μ-a.e. $\xi \in \mathbb{S}^{d-1}$. Since f has to fulfill the boundary condition $f(\pi,\xi) = 0$ for μ-a.e. $\xi \in \mathbb{S}^{d-1}$, there exists no function $f \in \mathcal{D}_0(\frac{\partial}{\partial t}; Z_\pi^d)$, $\|f\|_W = 1$, for which equality is attained in (2.29). $\quad\sqcap$

47

2.1.3. Inequalities in the non-compact case

We leave now the compact case and generalize the uncertainty principle of Theorem 1.32 to weighted L^2-spaces where the underlying domain is the d-dimensional one-sided tube

$$Z_\infty^d := \{(t,\xi): \, t \in [0,\infty), \, \xi \in \mathbb{S}^{d-1}\} \subset \mathbb{R}^{d+1}. \tag{2.30}$$

Similar as Z_π^d, also the non-compact set Z_∞^d is a Riemannian submanifold of $\mathbb{R}^{d+1}$ with boundary

$$\partial_L Z_\infty^d := \big\{(0,\xi): \, \xi \in \mathbb{S}^{d-1}\big\}. \tag{2.31}$$

The canonical measure on Z_∞^d is given by the product measure $dt d\mu(\xi)$.

Assumption 2.15. As admissible weight functions on Z_∞^d, we consider positive functions W satisfying the properties

$$(i) \qquad W \in C(Z_\infty^d): \quad W(\cdot,\xi) \in AC_{loc}([0,\infty)) \quad \text{for } \mu\text{-a.e. } \xi \in \mathbb{S}^{d-1},$$
$$W'|_K \in L^1(K) \quad \text{for all compact } K \in Z_\infty^d, \tag{2.32}$$

$$(ii) \qquad W(t,\xi) > 0 \quad \text{for all } (t,\xi) \in (0,\infty) \times \mathbb{S}^{d-1}, \tag{2.33}$$

$$(iii) \qquad t\frac{W'}{W} \in L^\infty(K) \quad \text{for all compact } K \in Z_\infty^d. \tag{2.34}$$

Assumption 2.15 can be considered as an adaption of Assumption 2.1 onto the non-compact tube Z_∞^d. The first condition (2.32) says that the weight function W is absolutely continuous with respect to the variable t for μ-a.e. fixed unit vector $\xi \in \mathbb{S}^{d-1}$, and that the Radon-Nikodym derivative $W' = \frac{\partial W}{\partial t}$ of W with respect to the variable t is a locally integrable function on Z_∞^d. The second property (2.33) implies that all the zeros of W are at the boundary $\partial_L Z_\infty^d$ of Z_∞^d. The third condition (2.34) guarantees that the fraction $t\frac{W'}{W}$ is essentially bounded on every compact subset of Z_∞^d. Further, we remark that there are no integrability restrictions on the weight function W. In fact, the integral $\int_{\mathbb{S}^{d-1}} \int_0^\infty W(t,\xi) d\mu(\xi) dt$ is not necessarily finite. This is, for instance, the case in the Heisenberg-Pauli-Weyl principle for $\mathbb{R}^d$ where the weight function W can be determined as $W(t,\xi) = t^{d-1}$ (see the upcoming Section 2.4).

Example 2.16. Consider the weight function $W_{\nu,r} : \, Z_\infty^d \to \mathbb{R}$, $W_{\nu,r}(t,\xi) = \sinh^\nu(rt)$, $\nu \geq 0$, $r > 0$. The conditions (2.32) and (2.33) are obviously satisfied. Moreover,

$$\left| t\frac{W_{\nu,r}'(t,\xi)}{W_{\nu,r}(t,\xi)} \right| = |r\nu t \coth(rt)| = \left| \nu \left(1 + \sum_{k=1}^\infty \frac{2r^2 t^2}{k^2(\pi^2 + r^2 t^2)} \right) \right| \leq \nu \left(1 + \frac{\pi^2}{3} \right).$$

Hence, also condition (2.34) is satisfied. In Section 2.6.4, we will see that for $\nu = d-1$ the weight function $W_{\nu,r}$ is related to the hyperbolic space $\mathbb{H}_r^d$.

Definition 2.17. For an admissible weight function W on Z_∞^d, we denote by $L^2(Z_\infty^d, W)$ the Hilbert space of weighted square integrable functions on Z_∞^d with the inner product

$$\langle f, g \rangle_W := \int_{\mathbb{S}^{d-1}} \int_0^\infty f(t,\xi)\overline{g(t,\xi)} W(t,\xi) dt d\mu(\xi) \tag{2.35}$$

As a substitute for the differential operator $\frac{\partial}{\partial t}$, we construct, similar as in the case of the nonnegative real half-axis (see (1.55)), a differential-difference operator on a symmetrically extended Hilbert space.

Definition 2.18. On the two-sided tube

$$Y^d := \mathbb{R} \times \mathbb{S}^{d-1} \subset \mathbb{R}^{d+1}, \tag{2.36}$$

we define the extended weight function $\tilde{W}$ by

$$\tilde{W}(t,\xi) := \frac{1}{2} W(|t|,\xi), \quad (t,\xi) \in Y^d. \tag{2.37}$$

By $L^2(Y^d, \tilde{W})$, we denote the Hilbert space of weighted square integrable functions on Y^d with the inner product

$$\langle f, g \rangle_{\tilde{W}} := \int_{\mathbb{S}^d} \int_{\mathbb{R}} f(t,\xi)\overline{g(t,\xi)} \tilde{W}(t,\xi) dt d\mu(\xi). \tag{2.38}$$

The set $Y^d \subset \mathbb{R}^{d+1}$ is a non-compact Riemannian manifold without boundary. The link between the spaces $L^2(Z_\infty^d, W)$ and $L^2(Y^d, \tilde{W})$ is given by the operators e and r:

$$e : L^2(Z_\infty^d, W) \to L^2(Y^d, \tilde{W}), \quad e(f)(t,\xi) := f(|t|,\xi) \quad \text{for a.e. } (t,\xi) \in Y^d, \tag{2.39}$$

$$r : L^2(Y^d, \tilde{W}) \to L^2(Z_\infty^d, W), \quad r(g) := g|_{[0,\infty) \times \mathbb{S}^{d-1}}. \tag{2.40}$$

If we define

$$L_e^2(Y^d, \tilde{W}) := \left\{ g \in L^2(Y^d, \tilde{W}) : g(t,\xi) = g(-t,\xi) \quad \text{for a.e. } (t,\xi) \in Y^d \right\} \tag{2.41}$$

as the subspace of even functions in $L^2(Y^d, \tilde{W})$, then the operators e and r constitute isometric isomorphisms between the Hilbert spaces $L^2(Z_\infty^d, W)$ and $L_e^2(Y^d, \tilde{W})$.

By property (2.32), the weight function W is continuous on Z_∞^d and its Radon-Nikodym derivative W' with respect to t is integrable on every compact subset K of Z_∞^d. Thus, also the even extension $\tilde{W}$ is continuous on Y^d, $\tilde{W}(\cdot,\xi) \in AC_{loc}(\mathbb{R})$ is locally absolutely continuous for μ-a.e. $\xi \in \mathbb{S}^{d-1}$ and the Radon-Nikodym derivative $\tilde{W}' = \frac{\partial \tilde{W}}{\partial t}$ is integrable on every compact subset K of the two-sided tube Y^d.

Definition 2.19. For a weight function W satisfying Assumption 2.15, we define the Dunkl operator T^Y on $L^2(Y^d, \tilde{W})$ as

$$T^Y g := \frac{\partial g}{\partial t} + \frac{\tilde{W}'}{\tilde{W}} \frac{g - \check{g}}{2}, \tag{2.42}$$

with the domain

$$\mathcal{D}(T^Y) := \left\{ g \in L^2(Y^d, \tilde{W}) : g(\cdot, \xi) \in AC_{loc}(\mathbb{R}) \quad \text{for } \mu\text{-a.e. } \xi \in \mathbb{S}^{d-1}, \right.$$
$$\left. \frac{\partial g}{\partial t}, \frac{\tilde{W}'}{\tilde{W}} \frac{g - \check{g}}{2} \in L^2(Y^d, \tilde{W}) \right\}, \tag{2.43}$$

where the reflection $\check{g}$ is defined by $\check{g}(t,\xi) = g(-t,\xi)$ for a.e. $(t,\xi) \in Y^d$.

Lemma 2.20.
The operator iT^Y with the domain $\mathcal{D}(T^Y)$ is symmetric and densely defined on $L^2(Y^d, \tilde{W})$.

Proof. We consider the subset $C_c^{1,t}(Y^d) := \left\{ g \in C_c(Y^d) : \frac{\partial g}{\partial t} \in C_c(Y^d) \right\}$ of the space $C_c(Y^d)$ of compactly supported and continuous functions on Y^d. $C_c^{1,t}(Y^d)$ is a subalgebra of $C_0(Y^d)$ that separates the points on Y^d, vanishes nowhere and is closed under complex conjugation (i.e. $\bar{g} \in C_c^{1,t}(Y^d)$ if $g \in C_c^{1,t}(Y^d)$). Then, by a variant of the Stone-Weierstrass Theorem (see Theorem B.2), every function $g_c \in C_c(Y^d) \subset C_0(Y^d)$ can be approximated uniformly by a function from $C_c^{1,t}(Y^d)$. Further, $C_c^{1,t}(Y^d)$ is a subset of $\mathcal{D}(T^Y)$ if W satisfies Assumption 2.15 (this follows in the same way as in Remark 2.7). Therefore, since the space $C_c(Y^d)$ is dense in $L^2(Y^d, \tilde{W})$ (see, for instance, [38, Theorem 13.21]), we can conclude that $\mathcal{D}(T^Y)$ is dense in $L^2(Y^d, \tilde{W})$.

For two functions f, g on $\mathcal{D}(T^Y)$, integration by parts with respect to the variable t is well defined (see equation (B.7)) and yields

$$\int_{\mathbb{S}^{d-1}} \int_{-\infty}^{\infty} \frac{\partial f}{\partial t}(t,\xi) g(t,\xi) \tilde{W}(t,\xi) dt d\mu(\xi) = -\int_{\mathbb{S}^{d-1}} \int_{-\infty}^{\infty} f(t,\xi) \frac{\partial}{\partial t} \left(g(t,\xi) \tilde{W}(t,\xi) \right) dt d\mu(\xi).$$

Now, following step by step the lines of the proof of Lemma 2.8 in which we proved the symmetry of the operator iT^X, we obtain the symmetry of iT^Y. $\qquad\square$

We define now the operators A and B on the Hilbert space $L^2(Y^d, \tilde{W})$ by

$$Ag(t,\xi) := t\, g(t,\xi), \quad \mathcal{D}(A) = \left\{ g \in L^2(Y^d, \tilde{W}) : tg \in L^2(Y^d, \tilde{W}) \right\}, \qquad (2.44)$$

$$Bg(t,\xi) := iT^Y g(t,\xi), \quad \mathcal{D}(B) = \mathcal{D}(T^Y). \qquad (2.45)$$

By Lemma 2.20, $B = iT^Y$ is symmetric, and for $f, g \in \mathcal{D}(A)$, we have

$$\langle Af, g \rangle_{\tilde{W}} = \int_{\mathbb{S}^{d-1}} \int_{\mathbb{R}} tf(t,\xi)\overline{g(t,\xi)} \tilde{W}(t,\xi) dt d\mu(\xi)$$

$$= \int_{\mathbb{S}^{d-1}} \int_{\mathbb{R}} f(t,\xi)\overline{tg(t,\xi)} \tilde{W}(t,\xi) dt d\mu(\xi) = \langle f, Ag \rangle_{\tilde{W}}.$$

Therefore, also A is symmetric. For functions f in

$$\mathcal{D}(\tfrac{\partial}{\partial t}, t, t\tfrac{\partial}{\partial t}; Z_\infty^d) := \left\{ f \in L^2(Z_\infty^d, W) : f(\cdot, \xi) \in AC_{loc}([0,\infty)) \quad \text{for } \mu\text{-a.e. } \xi \in \mathbb{S}^{d-1}, \right.$$

$$\left. tf, \frac{\partial f}{\partial t}, t\frac{\partial f}{\partial t}, t\frac{W'}{W} f \in L^2(Z_\infty^d, W) \right\}, \qquad (2.46)$$

we get now the following inequality:

Theorem 2.21.
Suppose that the weight function W satisfies Assumption 2.15. Let $f \in L^2(Z_\infty^d, W) \cap \mathcal{D}(\tfrac{\partial}{\partial t}, t, t\tfrac{\partial}{\partial t}; Z_\infty^d)$ such that $\|f\|_W = 1$. Then, the following inequality holds:

$$\|tf\|_W^2 \cdot \left\| \frac{\partial f}{\partial t} \right\|_W^2 \geq \frac{1}{4} \left| 1 + \int_{\mathbb{S}^{d-1}} \int_0^\infty tW'(t,\xi)|f(t,\xi)|^2 dt d\mu(\xi) \right|^2. \qquad (2.47)$$

Equality in (2.47) is attained if and only if $f(t,\xi) = C(\xi)e^{-\lambda t^2}$, with a complex valued function $C : \mathbb{S}^{d-1} \to \mathbb{C}$ and a real constant $\lambda \in \mathbb{R}$ which have to be chosen such that f satisfies the requirements of the theorem.

Proof. We consider the operators A and B defined in (2.44) and (2.45) on the Hilbert space $L^2(Y^d, \tilde{W})$. The commutator of A and B is given by

$$[A,B]g = it\frac{\partial g}{\partial t} + it\frac{\tilde{W}'}{\tilde{W}}\frac{g - \check{g}}{2} - it\frac{\partial g}{\partial t} - ig(t,\xi) - it\frac{\tilde{W}'}{\tilde{W}}\frac{g + \check{g}}{2}, \qquad (2.48)$$

for all functions $g \in \mathcal{D}(AB) \cap \mathcal{D}(BA)$. If we consider only even functions $g \in L^2_e(Y^d, \tilde{W})$, we get

$$[A,B]g = -ig\left(1 + t\frac{\tilde{W}'}{\tilde{W}}\right).$$

Next, if $f \in L^2(Z^d_\infty, W) \cap \mathcal{D}(\frac{\partial}{\partial t}, t, t\frac{\partial}{\partial t}; Z^d_\infty)$, then the even extension $e(f) \in L^2(Y^d, \tilde{W})$ is an element of the domain $\mathcal{D}([A,B]) = \mathcal{D}(AB) \cap \mathcal{D}(BA)$. Hence, if we apply the symmetric operators A and B to the even function $e(f)$, we obtain in inequality (1.4):

$$\|Ae(f)\|^2_{\tilde{W}} = \|te(f)\|^2_{\tilde{W}} = \|tf\|^2_W,$$
$$a = \langle Ae(f), e(f)\rangle_{\tilde{W}} = \langle te(f), e(f)\rangle_{\tilde{W}} = 0,$$
$$\|Be(f)\|^2_{\tilde{W}} = \|iT^Y e(f)\|^2_{\tilde{W}} = \left\|\frac{\partial e(f)}{\partial t}\right\|^2_{\tilde{W}} = \left\|\frac{\partial f}{\partial t}\right\|^2_W,$$
$$b = \langle Be(f), e(f)\rangle_{\tilde{W}} = \langle i\frac{\partial e(f)}{\partial t}, e(f)\rangle_{\tilde{W}} = 0,$$
$$[A,B]e(f) = -ie(f)\left(1 + t\frac{\tilde{W}'}{\tilde{W}}\right).$$

Further, since the identity map $t \to t$ and the function $\tilde{W}'$ are odd in t and f is a normalized function, we conclude

$$|\langle [A,B]e(f), e(f)\rangle_{\tilde{W}}| = \left|\left\langle \left(1 + t\frac{\tilde{W}'}{\tilde{W}}\right)e(f), e(f)\right\rangle_{\tilde{W}}\right|$$
$$= \left|1 + \int_{\mathbb{S}^{d-1}} \int_0^\infty tW'(t,\xi)|f(t,\xi)|^2 dt\right|.$$

Finally, by Theorem 1.2, equality in (2.47) is attained if and only if

$$i\frac{\partial e(f)}{\partial t} = -i2\lambda te(f),$$

where λ denotes a real constant. The solution of this differential equation corresponds to a function $e(f)$ of the form $e(f)(t,\xi) = C(\xi)e^{-\lambda t^2}$. Restricted to the nonnegative half-part Z^d_∞ of the tube Y^d, this yields the assertion. $\qquad \square$

Example 2.22. For the weight function $W_{\nu,r}(t,\xi) = \sinh^\nu(rt)$, $\nu \geq 0$, $r > 0$, the Dunkl operator T^Y on $L^2(Y^d, \tilde{W}_{\nu,r})$ reads as

$$T^Y g(t,\xi) = \frac{\partial g}{\partial t}(t,\xi) + \nu\, r \coth(rt) \frac{g(t,\xi) - g(-t,\xi)}{2}, \quad \text{for a.e. } (t,\xi) \in Y^d.$$

Then, inequality (2.47) attains the form

$$\|tf\|_{W_{\nu,r}}^2 \cdot \left\|\frac{\partial f}{\partial t}\right\|_{W_{\nu,r}}^2 \geq \frac{1}{4}\left|1 + \nu\, r \int_{\mathbb{S}^{d-1}} \int_0^\infty t|f(t,\xi)|^2 \cosh^{\nu-1}(rt)dt d\mu(\xi)\right|^2. \tag{2.49}$$

2.2. Uncertainty principles on compact Riemannian manifolds

This section is one of the main parts of this work. In the following, we will combine the weighted L^2-inequalities developed in Section 2.1.1 with the geometry of a compact Riemannian manifold M and derive an uncertainty principle for functions on $L^2(M)$. In a first step, we will show that the geometric structure of the compact manifold M leads to an isometric isomorphism between the Hilbert space $L^2(M)$ and a weighted L^2-space $L^2(Z_\pi^d, W_{M,p})$ on the cylindrical domain Z_π^d. Then, we will use the uncertainty inequality (2.24) for the space $L^2(Z_\pi^d, W_{M,p})$ to prove an uncertainty principle for compact Riemannian manifolds.

2.2.1. An isomorphism between $L^2(M)$ and $L^2(Z_\pi^d, W_{M,p})$

In this first part, we will show that the Hilbert space $L^2(M)$ of square integrable functions on a compact Riemannian manifold M is isometrically isomorphic to a weighted L^2-space $L^2(Z_\pi^d, W_{M,p})$ on the domain Z_π^d. To this end, we will construct a chart that maps the domain Z_π^d onto the compact manifold M and a pull back operator that maps functions on M to functions on Z_π^d. In total, we need three mappings: the exponential map $\exp_p$, the polar transform P and a further coordinate transform L_R.

We start out by recapitulating some basics and refer to the Appendix A for a short introduction into Riemannian manifolds. Over the entire section, we denote by M a simply connected, d-dimensional, compact Riemannian manifold without boundary and by T_pM the tangent space at the point $p \in M$. Further, we denote by μ_M the canonical Riemannian measure on M (see Section A.5).

Definition 2.23. We define the Hilbert space $L^2(M)$ of square integrable functions on M as

$$L^2(M) := \left\{ f : M \to \mathbb{C} : f \text{ Borel measurable}, \int_M |f(q)|^2 d\mu_M(q) < \infty \right\} \tag{2.50}$$

with the inner product

$$\langle f, g \rangle_M := \int_M f(q)\overline{g(q)} d\mu_M(q), \quad f, g \in L^2(M), \tag{2.51}$$

and the norm $\|f\|_M := \sqrt{\langle f, f \rangle_M}$, with the usual understanding that two functions $f, g \in L^2(M)$ are identified with each other if $f(q) = g(q)$ for μ_M-a.e. $q \in M$, i.e., for all $q \in M$ except a set of μ_M-measure zero.

The proof that the Hilbert space $L^2(M)$ is well-defined and complete in the topology induced by the norm $\| \cdot \|_M$ is standard. For further details, we refer to Section B.1 of the appendix.

Definition 2.24. A distance metric $d(p, q)$ between two points p and q on the manifold M is defined by

$$d(p, q) := \inf_\gamma \int_a^b |\gamma'(t)| dt, \tag{2.52}$$

where γ ranges over all piecewise differentiable paths $\gamma : [a, b] \to M$ satisfying $\gamma(a) = p$ and $\gamma(b) = q$. For $p \in M$ and $\delta > 0$, we introduce on M the open balls and spheres with center p as

$$B(p, \delta) := \{q \in M, \ d(q, p) < \delta\}, \tag{2.53}$$
$$S(p, \delta) := \{q \in M, \ d(q, p) = \delta\}. \tag{2.54}$$

Similarly, we define on the tangent space $T_p M$

$$\mathfrak{B}(p, \delta) := \{\xi \in T_p M, \ |\xi| < \delta\}, \tag{2.55}$$
$$\mathfrak{S}(p, \delta) := \{\xi \in T_p M, \ |\xi| = \delta\}, \tag{2.56}$$
$$\mathfrak{S}_p := \mathfrak{S}(p, 1), \tag{2.57}$$

where $|\xi|$ denotes the Euclidean length of ξ in the tangent space $T_p M$.

Now, as a first step to get a mapping from Z_π^d onto M, we consider the exponential map $\exp_p$ from the tangent space $T_p M$ onto M.

Definition 2.25. Let $p \in M$ be fixed, $\xi \in T_p M$ and $\gamma_\xi : \mathbb{R} \to M$ be the locally unique geodesic with initial conditions $\gamma_\xi(0) = p$ and $\gamma'(0) = \xi$ (see Section A.3). Then, the exponential map $\exp_p : T_p M \to M$ is defined as

$$\exp_p(\xi) := \gamma_\xi(1). \tag{2.58}$$

Since M is compact and, hence, topologically complete, the Theorem of Hopf and Rinow (Theorem A.2) ensures that the geodesic γ_ξ can be defined on the whole real line $\mathbb{R}$. Thus, also the exponential map is well-defined on the whole tangent space $T_p M$. The exponential map $\exp_p$ defines a local diffeomorphism from a neighborhood of the origin 0 in the tangent space T_pM onto a neighborhood of p in M. In particular, for a small $\delta > 0$, the exponential $\exp_p$ maps the balls $\mathfrak{B}(p, \delta) \subset T_pM$ isometrically onto the balls $B(p, \delta) \subset M$, i.e.,

$$\exp_p \mathfrak{B}(p, \delta) = B(p, \delta),$$
$$\exp_p \mathfrak{S}(p, \delta) = S(p, \delta).$$

Moreover, there exists a maximal star-shaped open domain $\mathfrak{D}_p$ of the tangent space T_pM for which the exponential map $\exp_p$ is a diffeomorphism and for which the image $D_p = \exp_p \mathfrak{D}_p$ covers the whole manifold M up to a set C_p of Riemannian measure zero (see Theorem A.4), i.e.,

$$M = D_p \cup C_p. \tag{2.59}$$

The null set C_p corresponds to the image $\exp_p \mathfrak{C}_p$ of the boundary $\mathfrak{C}_p = \partial \mathfrak{D}_p$ of $\mathfrak{D}_p$ and is called the cut locus of p. Due to (A.30), we have the following formula for integrable functions f on M:

$$\int_M f(q) d\mu_M(q) = \int_{\overline{\mathfrak{D}_p}} f(\exp_p(\xi))\theta(\xi) d\xi, \tag{2.60}$$

where $\theta(\xi) := \det((d\exp_p)_\xi)$ denotes the Jacobian determinant of the exponential map $\exp_p$ and $\overline{\mathfrak{D}_p} = \mathfrak{D}_p \cup \mathfrak{C}_p$.

Example 2.26. For the unit sphere $\mathbb{S}^2 = \{q \in \mathbb{R}^3 : |q|^2 = 1\}$, the tangent space $T_p\mathbb{S}^2$ of a point $p \in \mathbb{S}^2$ can be identified with the orthogonal complement $p^\perp$ of the linear vector space $\mathbb{R}p$ in $\mathbb{R}^3$. The cut locus C_p of p consists of the antipodal point $\{-p\}$ and $D_p = \mathbb{S}^2 \setminus \{-p\}$. The geodesics γ_ξ through the point p correspond to the great circles passing through p. Further, $\mathfrak{D}_p = \mathfrak{B}(p, \pi)$, $\mathfrak{C}_p = \mathfrak{S}(p, \pi)$ and the Jacobian determinant $\theta(\xi)$ can be computed as (cf. [3, p. 57])

$$\theta(\xi) = \frac{\sin(|\xi|)}{|\xi|}.$$

Hence, by (2.60), we get for integrable functions f on $\mathbb{S}^2$ the formula

$$\int_{\mathbb{S}^2} f(q) d\mu_{\mathbb{S}^2}(q) = \int_{\overline{\mathfrak{B}(p,\pi)}} f(\exp_p(\xi)) \tfrac{\sin(|\xi|)}{|\xi|} d\xi. \tag{2.61}$$

A draft of the exponential map $\exp_p$ on $T_p\mathbb{S}^2$ can be seen in Figure 3.

Next, we are going to introduce polar coordinates on the tangent space T_pM. For a precise distinction between the settings, we will always use the symbol $\mathfrak{S}_p$ to denote the $(d-1)$-dimensional unit sphere in the tangent space T_pM, and the symbol $\mathbb{S}^{d-1}$ to denote the unit sphere in $\mathbb{R}^d$.

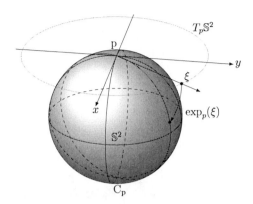

Figure 3: The exponential map $\exp_p$ on the unit sphere $\mathbb{S}^2$.

Definition 2.27. On the tubes $[0, \infty) \times \mathfrak{S}_p$ and $Z_\infty^d = [0, \infty) \times \mathbb{S}^{d-1}$, we define the polar transform P by

$$\text{P} : [0, \infty) \times \mathfrak{S}_p \to T_p M : \quad \text{P}(t, \xi) = t\xi, \tag{2.62}$$

$$\text{P} : Z_\infty^d \to \mathbb{R}^d : \quad \text{P}(t, \xi) = t\xi, \tag{2.63}$$

and the inverse polar transform by

$$\text{P}^{-1} : T_p M \setminus \{0\} \to (0, \infty) \times \mathfrak{S}_p : \quad \text{P}^{-1}(\xi) = (|\xi|, \tfrac{1}{|\xi|}\xi), \tag{2.64}$$

$$\text{P}^{-1} : \mathbb{R}^d \setminus \{0\} \to Z_\infty^d \setminus \partial_L Z_\infty^d : \quad \text{P}^{-1}(\xi) = (|\xi|, \tfrac{1}{|\xi|}\xi). \tag{2.65}$$

Now, using the polar transform P, we want to describe the set $\mathfrak{D}_p \subset T_p M$ in terms of the coordinates $(t, \xi) \in [0, \infty) \times \mathfrak{S}_p$. To this end, we define

$$R(\xi) := \sup_{t>0} \{t\xi \in \mathfrak{D}_p\}, \quad \xi \in \mathfrak{S}_p, \tag{2.66}$$

as the Euclidean distance from the origin to the boundary $\mathfrak{C}_p = \partial \mathfrak{D}_p$ in direction $\xi \in \mathfrak{S}_p$ (for an equivalent definition see also (A.18)). The distance R, considered as a real-valued function on $\mathfrak{S}_p$, is strictly positive and, moreover, Lipschitz continuous on $\mathfrak{S}_p$ (see Theorem A.3 (c)). With help of the distance function R, we can define the pre-image of the set $\overline{\mathfrak{D}_p}$ under the polar transform P.

Definition 2.28. We define the d-dimensional subset Z_R^d of $[0, \infty) \times \mathfrak{S}_p$ as

$$Z_R^d := \{(t, \xi) : \ t \in [0, R(\xi)], \ \xi \in \mathfrak{S}_p\} \subset [0, \infty) \times \mathfrak{S}_p, \tag{2.67}$$

with the boundary

$$\partial_L Z_R^d := \{(0, \xi) : \ \xi \in \mathfrak{S}_p\}, \tag{2.68}$$

$$\partial_R Z_R^d := \{(R(\xi), \xi) : \ \xi \in \mathfrak{S}_p\}. \tag{2.69}$$

The set Z_R^d is clearly a compact subset of $[0, \infty) \times \mathfrak{S}_p$ and the polar transform P maps Z_R^d onto $\overline{\mathfrak{D}_p}$. Moreover, the transformation P defines a diffeomorphism from $Z_R^d \setminus \{\partial_L Z_R^d, \partial_R Z_R^d\}$ onto the open set $\mathfrak{D}_p \setminus \{0\}$. Combining the polar transform P with the exponential map $\exp_p$, we have

$$\exp_p P(Z_R^d) = M,$$

and $\exp_p P$ defines a diffeomorphism from $Z_R^d \setminus \{\partial_L Z_R^d, \partial_R Z_R^d\}$ onto $D_p \setminus \{p\}$. Hence, the points $(t, \xi) \in Z_R^d$ determine a coordinate system on the manifold M and are usually referred to as geodesic polar coordinates on M. Further, the polar transform P induces a change of variables that yields the formula (see [75, Theorem 8.26])

$$\int_{\overline{\mathfrak{D}_p}} f(\exp_p(\xi))\theta(\xi)d\xi = \int_{\mathfrak{S}_p} \int_0^{R(\xi)} f(\exp_p(t\xi))t^{d-1}\theta(t\xi)dtd\mu(\xi), \qquad (2.70)$$

where μ denotes the standard Riemannian measure on the unit sphere $\mathfrak{S}_p$ and the term t^{d-1} corresponds to the Jacobian determinant of P.

To simplify the notation, we introduce the following pull backs of a function f on M:

Definition 2.29. For a function $f : M \to \mathbb{C}$, we define the pull back functions

$$\exp_p^* f : \overline{\mathfrak{D}_p} \to \mathbb{C}, \quad \exp_p^* f(\xi) := f(\exp_p(\xi)), \qquad (2.71)$$
$$f^* : Z_R^d \to \mathbb{C}, \quad f^*(t, \xi) := P^* \exp_p^* f(t, \xi) := \exp_p^* f(t\xi) = f(\exp_p(t\xi)). \qquad (2.72)$$

The original function f and the pull backs f^* and $\exp_p^* f$ are related by the following commutative diagram:

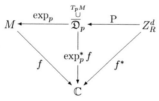

Figure 4: The relation between the functions f, f^* and $\exp_p^* f$.

On Z_R^d, we introduce the weight function Θ by

$$\Theta(t, \xi) := t^{d-1}\theta(t\xi). \qquad (2.73)$$

Then, formula (2.70) can be rewritten as

$$\int_{\overline{\mathfrak{D}_p}} \exp_p^* f(\xi)\theta(\xi)d\xi = \int_{\mathfrak{S}_p} \int_0^{R(\xi)} f^*(t, \xi)\Theta(t, \xi)dt\, d\mu(\xi). \qquad (2.74)$$

Example 2.30. For the unit sphere $\mathbb{S}^2$, the distance $R(\xi)$ to the cut locus is, independently of the directional variable $\xi \in \mathfrak{S}_p$, always equal to π. Hence, if we parameterize the one-dimensional unit sphere $\mathfrak{S}_p$ using an angular variable $\varphi \in [0, 2\pi)$, the formulas (2.61) and (2.74) imply the following identity:

$$\int_{\mathbb{S}^2} f(q) d\mu_{\mathbb{S}^2}(q) = \int_0^{2\pi} \int_0^\pi f^*(t, \varphi) \sin t \, dt \, d\varphi. \qquad (2.75)$$

Up to now, we have built a coordinate transformation $\exp_p \mathrm{P}$ that maps the set Z_R^d onto M. To get the desired map from the cylinder Z_π^d onto M, we have to introduce a third coordinate transform form Z_π^d onto Z_R^d.

Definition 2.31. On Z_π^d, we define the coordinate transform L_R as

$$\mathrm{L}_R : Z_\pi^d \to Z_R^d, \quad (\tau, \xi) \to (\tfrac{R(\xi)}{\pi} \tau, \xi). \qquad (2.76)$$

The mapping $\mathbb{S}^{d-1} \to \mathfrak{S}_p$, $\xi \to \xi$, in (2.76) is well defined in the sense that there exists a canonical identification between a unit vector $\xi \in \mathbb{S}^{d-1} \subset \mathbb{R}^d$ and an element $\xi \in \mathfrak{S}_p \subset T_pM$. Since the function $\xi \to R(\xi)$ is strictly positive and Lipschitz continuous on $\mathfrak{S}_p$, also the inverse function $\xi \to \frac{1}{R(\xi)}$ is strictly positive and Lipschitz continuous on $\mathfrak{S}_p$ (for the definition of Lipschitz continuous, see Section B.1). Hence, the mapping L_R defines a lipeomorphism (or a bi-Lipschitzian mapping) from Z_π^d onto Z_R^d. Moreover, the points $(\tau, \xi) \in Z_\pi^d$ form a new coordinate system for the compact manifold M.

For the integral of a function over a domain, a lipeomorphism yields a similar transformation formula as a diffeomorphism (see [88, Section 2.2]). In our case, the lipeomorphism L_R induces a change of variables that yields the following identity:

$$\int_{\mathfrak{S}_p} \int_0^{R(\xi)} f^*(t, \xi) \Theta(t, \xi) dt d\mu(\xi) = \int_{\mathbb{S}^{d-1}} \int_0^\pi f^*(\tfrac{R(\xi)}{\pi}\tau, \xi) \tfrac{\pi}{R(\xi)} \Theta(\tfrac{R(\xi)}{\pi}\tau, \xi) d\tau d\mu(\xi), \qquad (2.77)$$

where the Jacobian determinant of L_R equals $\frac{\pi}{R}$ almost everywhere.

Definition 2.32. For a function f^* on Z_R^d, we define the pull back by the lipeomorphism L_R as

$$\mathrm{L}_R^* f^* : Z_\pi^d \to \mathbb{C}, \quad \mathrm{L}_R^* f^*(\tau, \xi) := f^*(\tfrac{R(\xi)}{\pi}\tau, \xi), \quad (\tau, \xi) \in Z_\pi^d, \qquad (2.78)$$

and introduce the weight function $W_{M,p}$ on Z_π^d as

$$W_{M,p}(\tau, \xi) := \tfrac{\pi}{R(\xi)} \Theta(\tfrac{R(\xi)}{\pi}\tau, \xi), \quad (\tau, \xi) \in [0, \pi] \times \mathbb{S}^{d-1}. \qquad (2.79)$$

Example 2.33. Let M be the two-dimensional quadratic flat torus $\mathbb{T}_\pi^2 = \mathbb{R}^2/(2\pi\mathbb{Z})^2$ with side length 2π. For any point $p \in \mathbb{T}_\pi^2$, the set $\mathfrak{D}_p$ in the tangent space $T_p\mathbb{T}_\pi^2$ consists of the points in the open square

$$\square_\pi = \{(\xi_1, \xi_2) \in T_p\mathbb{T}_\pi^2 : -\pi < \xi_1, \xi_2 < \pi\}$$

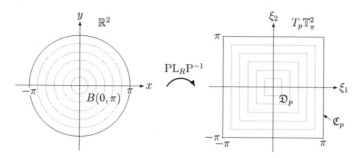

Figure 5: The lipeomorphism $\mathrm{PL}_R\mathrm{P}^{-1}$ in the case that $M = \mathbb{T}^2_\pi$ is the two-dimensional flat torus with side length 2π.

and the tangential cut locus $\mathfrak{C}_p$ corresponds exactly with the boundary of $\square_\pi$. Then, $\mathrm{PL}_R\mathrm{P}^{-1}$ maps the balls $B(0,r) \setminus \{0\}$ with radius $r > 0$ in $\mathbb{R}^2$ onto the squares $\square_r \setminus \{0\}$ with side length $2r$ (centered at the origin) in $T_p\mathbb{T}^2_\pi$, see Figure 5. This example will be further discussed in Section 2.6.3.

Proposition 2.34.
The Hilbert space $L^2(M)$ is isometrically isomorphic to the Hilbert space $L^2(Z^d_\pi, W_{M,p})$ with the weight function $W_{M,p}$ defined in (2.79). The isomorphism is explicitly given by the pull back operator $\mathrm{L}^*_R\mathrm{P}^*\exp^*_p$.

Proof. For the proof of Proposition 2.34, we just have to collect the changes of variables induced by the exponential map $\exp_p$, the polar transform P and the lipeomorphism L_R. If f, g in $L^2(M)$, then we get

$$
\begin{aligned}
\langle f, g \rangle_M &\overset{(2.51)}{=} \int_M f(q)\overline{g(q)}d\mu_M(q) \\
&\overset{(2.60)}{=} \int_{\widehat{\mathfrak{D}}_p} f(\exp_p(\xi))\overline{g(\exp_p(\xi))}\theta(\xi)d\xi \\
&\overset{(2.74)}{=} \int_{\mathfrak{S}_p} \int_0^{R(\xi)} f^*(t,\xi)\overline{g^*(t,\xi)}\Theta(t,\xi)dt d\mu(\xi) \\
&\overset{(2.77)}{=} \int_{\mathbb{S}^{d-1}} \int_0^\pi f^*(\tfrac{R(\xi)}{\pi}\tau,\xi)\overline{g^*(\tfrac{R(\xi)}{\pi}\tau,\xi)}\tfrac{\pi}{R(\xi)}\Theta(\tfrac{R(\xi)}{\pi}\tau,\xi)d\tau d\mu(\xi) \\
&\overset{(2.78)}{\underset{(2.79)}{=}} \int_{\mathbb{S}^{d-1}} \int_0^\pi \mathrm{L}^*_Rf^*(\tau,\xi)\overline{\mathrm{L}^*_Rg^*(\tau,\xi)}W_{M,p}(\tau,\xi)d\tau d\mu(\xi) \\
&\overset{(2.7)}{=} \langle \mathrm{L}^*_Rf^*, \mathrm{L}^*_Rg^* \rangle_{W_{M,p}}.
\end{aligned}
$$

Hence, the pull back operator $\mathrm{L}^*_R\mathrm{P}^*\exp^*_p$ is an isometric homomorphism from $L^2(M)$ to $L^2(Z^d_\pi, W_{M,p})$. Since the point set $\{p\}$ and the cut locus C_p are subsets of μ_M-measure zero in M and $\exp_p \mathrm{PL}_R$ defines a lipeomorphism from $Z^d_\pi \setminus \{\partial_L Z^d_\pi, \partial_R Z^d_\pi\}$ onto $M \setminus \{p, C_p\}$, we can conclude that the operator $L_R^* P^* \exp_p^* : L^2(M) \to L^2(Z_\pi^d, W_{M,p})$ is also surjective and, thus, an isomorphism. □

Remark 2.35. In the proof of Proposition 2.34 is implicitly stated that the Hilbert spaces $L^2(M)$ and $L^2(Z_\pi^d, W_{M,p})$ are also isometrically isomorphic to the Hilbert space $L^2(\overline{\mathfrak{D}}_p, \theta)$ with inner product

$$\langle \varphi_1, \varphi_2 \rangle_\theta := \int_{\mathfrak{D}_p} \varphi_1(\xi) \overline{\varphi_2(\xi)} \theta(\xi) d\xi$$

and to the Hilbert space $L^2(Z_R^d, \Theta)$ with inner product

$$\langle \psi_1, \psi_2 \rangle_\Theta := \int_{\mathfrak{S}_p} \int_0^{R(\xi)} \psi_1(t, \xi) \overline{\psi_2(t, \xi)} \Theta(t, \xi) dt d\mu(\xi).$$

A summary for the links between the different Hilbert spaces is given in Figure 6.

$$M \xleftarrow{\exp_p} \begin{array}{c} \overline{\mathfrak{D}}_p \\ \cup \\ T_p M \end{array} \xleftarrow{\text{P}} Z_R^d \xleftarrow{L_R} Z_\pi^d$$

$$L^2(M) \xrightarrow{\exp_p^*} L^2(\overline{\mathfrak{D}}_p, \theta) \xrightarrow{\text{P}^*} L^2(Z_R^d, \Theta) \xrightarrow{L_R^*} L^2(Z_\pi^d, W_{M,p})$$

Figure 6: The mappings $\exp_p$, P and L_R and the respective pull backs.

2.2.2. Uncertainty principles on compact Riemannian manifolds

The goal of this section is to prove an uncertainty principle for functions f in the Hilbert space $L^2(M)$. To this end, we will use the isomorphism $L_R^* P^* \exp_p^* : L^2(M) \to L^2(Z_\pi^d, W_{M,p})$ established in the last section, and adopt then Theorem 2.10 to get the desired uncertainty inequality. First, we will show that the weight function $W_{M,p}$ defined in (2.79) is admissible on Z_π^d, i.e. that it satisfies Assumption 2.1.

Lemma 2.36.
The weight function $W_{M,p}$ on Z_π^d satisfies Assumption 2.1.

Proof. The Jacobi determinant of the exponential map

$$\theta(t\xi) = \det((d \exp_p)_{g\xi}), \quad (t, \xi) \in [0, \infty) \times \mathfrak{S}_p,$$

is a positive and continuously differentiable function on $\overline{\mathfrak{D}}_p$ and the zeros of θ, called the conjugate points of p, lie at the boundary $\mathfrak{C}_p$ of $\overline{\mathfrak{D}}_p$ (see [12, XII, Proposition 2.2] and [8, Theorem II.5.5]). Since the distance function R is Lipschitz continuous on $\mathfrak{S}_p$ (cf. Theorem A.3 (c)) the weight function $W_{M,p}$ given by

$$W_{M,p}(\tau, \xi) = \frac{R(\xi)^{d-2}}{\pi^{\frac{d}{2}}} \tau^{d-1} \theta\left(\frac{R(\xi)}{\pi} \tau \xi\right)$$

and its derivative $W'_{M,p} = \frac{\partial W_{M,p}}{\partial \tau}$ with respect to the variable τ are continuous functions on Z^d_π. Hence, conditions (2.4) and (2.5) are satisfied.

By construction, the zeros of $W_{M,p}$ lie at the boundary $\partial_L Z^d_\pi$ and $\partial_R Z^d_\pi$ of the cylinder Z^d_π. Therefore, to validate property (2.6) we have to check that the function $\tau(\pi - \tau)\frac{W'}{W}$ is bounded at the boundary of Z^d_π. Due to [49, Chapter X, Corollary 3.3], $\theta(t\xi)$ has the following Taylor expansion at $t = 0$:

$$\theta(t\xi) = 1 - \frac{\mathrm{Ric}(\xi,\xi)}{6}t^2 + \mathrm{O}(t^3). \qquad (2.80)$$

Therefore, the weight function $W_{M,p}(\tau,\xi)$ has the following Taylor expansion at $\tau = 0$:

$$W_{M,p}(\tau,\xi) = \frac{R(\xi)^{d-2}}{\pi^{d-2}}\tau^{d-1} - \frac{R(\xi)^d}{\pi^d}\frac{\mathrm{Ric}(\xi,\xi)}{6}\tau^{d+1} + \mathrm{O}(\tau^{d+2}).$$

Thus,

$$\tau\frac{W'_{M,p}(\tau,\xi)}{W_{M,p}(\tau,\xi)} = (d-1) + \mathrm{O}(\tau^2)$$

and $\tau(\pi - \tau)W'_{M,p}/W_{M,p}$ is bounded in a small neighborhood at $\tau = 0$. Similarly, a Taylor expansion of $\theta(t\xi)$ at $t = R(\xi)$ (using Jacobi vector fields, see [49], Propositions IX.5.1, IX.5.3 and Proposition X.3.1) can be computed as

$$\theta(t\xi) = c_\xi(R(\xi) - t)^k + \mathrm{O}((R(\xi) - t)^{k+1}),$$

where $c_\xi > 0$ and $0 \leq k \leq d - 1$ is the dimension of the kernel of the Jacobi matrix $(d\exp_p)_{R(\xi)\xi}$. So, the fraction

$$\tau(\pi - \tau)\frac{W'_{M,p}(\tau,\xi)}{W_{M,p}(\tau,\xi)} = -k\pi + \mathrm{O}(\pi - \tau), \quad \tau \to \pi,$$

is also bounded at the right hand boundary $\partial_R Z^d_\pi$ of Z^d_π. In total, we can conclude that $\tau(\pi - \tau)\frac{W'_{M,p}(\tau,\xi)}{W_{M,p}(\tau,\xi)}$ is uniformly bounded on Z^d_π and, hence, that property (2.6) is satisfied.
$\square$

In principal, Theorem 2.10 can now be adopted to determine an uncertainty principle for functions on a compact Riemannian manifold. Beforehand, however, we will discuss the mapping $\exp_p \mathrm{PL}_R : Z^d_\pi \to M$ and the relation between continuous functions on M and Z^d_π in more detail. Further, we will investigate how the Dunkl operator on the space $L^2(X^d, \tilde{W}_{M,p})$ is related to a differential operator on M.

The composition $\exp_p \mathrm{PL}_R$ is a continuous mapping from Z^d_π onto M. Moreover, $\exp_p \mathrm{PL}_R$ maps the left hand boundary $\partial_L Z^d_\pi$ of the cylinder Z^d_π onto the point p and the right hand boundary $\partial_R Z^d_\pi$ onto the cut locus C_p of p. Hence, the image $\mathrm{L}^*_R f^*$ of a continuous function f on M is also continuous on the cylinder Z^d_π, but not every function $g \in C(Z^d_\pi)$

is the image of a continuous function on M under the pull back operator $L_R^* P^* \exp_p^*$. This is, for instance, the case if $g \in C(Z_\pi^d)$ satisfies $g(0, \xi_1) \neq g(0, \xi_2)$ for $\xi_1 \neq \xi_2$. So, if we want to identify a continuous function g on Z_π^d with a continuous function f on M, the function g has to satisfy additional consistency conditions. These consistency conditions are related to the topology of the compact Riemannian manifold M and, in particular, to the form of the cut locus C_p.

Definition 2.37. A continuous function g on Z_π^d is called topologically consistent with a continuous function f on M under the mapping $\exp_p \mathrm{PL}_R$ if $g(t, \xi) = L_R^* f^*(t, \xi)$ for all $(t, \xi) \in Z_\pi^d$. In this case, g satisfies the following consistency conditions:

$$g(0, \xi_1) = g(0, \xi_2) \quad \text{for all } \xi_1, \xi_2 \in \mathbb{S}^{d-1}, \tag{2.81}$$

$$g(\pi, \xi_1) = g(\pi, \xi_2) \quad \text{if } \exp_p \mathrm{PL}_R(\pi, \xi_1) = \exp_p \mathrm{PL}_R(\pi, \xi_2) \text{ on } M. \tag{2.82}$$

Moreover, we introduce the function space C^M as

$$C^M(Z_\pi^d) := \left\{ g \in C(Z_\pi^d) : \ g \text{ satisfies (2.81) and (2.82)} \right\}. \tag{2.83}$$

The condition (2.81) implies that g is constant at the left hand boundary $\partial_L Z_\pi^d$ of Z_π^d. Further, if the function g on Z_π^d is topologically consistent with $f \in C(M)$, then the constant value at $\partial_L Z_\pi^d$ corresponds exactly with the value $f(p)$. The second condition (2.82) ensures the topological consistency of the function g with the function f at the points of the cut locus C_p of p, i.e., if $\exp_p \mathrm{PL}_R$ maps two different points $(\pi, \xi_1), (\pi, \xi_2) \in Z_\pi^d$ onto the same point $q \in C_p$, then $g(\pi, \xi_1) = g(\pi, \xi_2) = f(q)$. Moreover, the operator $L_R^* P^* \exp_p^*$ defines an isometric isomorphism from the space $C(M)$ onto the space $C^M(Z_\pi^d)$ in the uniform norm.

In the following, whenever we consider functions on Z_π^d that are supposed to reflect the topological structure of the Riemannian manifold M we will ensure that the conditions (2.81) and (2.82) are satisfied. In particular, these conditions will be added in the upcoming definition of the domain of the radial differential operator.

Next, we want to introduce a radial frequency variance on the compact Riemannian manifold M and relate it to a Dunkl operator T^X. First of all, we introduce a new notation for operators that enables us to switch easily between operators described in geodesic polar coordinates $(t, \xi) \in Z_R^d$ and operators on $L^2(M)$.

Definition 2.38. For an operator A on the Hilbert space $L^2(Z_R^d, \Theta)$, we define its counterpart A_* on $L^2(M)$ by

$$A_* : L^2(M) \to L^2(M) : \quad (A_* f)^* := A f^*. \tag{2.84}$$

In particular, we define the multiplication with a function $h \in L^2(Z_R^d, \Theta)$ by

$$(h_* f)^*(t, \xi) := h(t, \xi) f^*(t, \xi) \quad \text{for a.c. } (t, \xi) \in Z_R^d,$$

and the radial differential operator $\frac{\partial}{\partial t_*}$ with respect to $p \in M$ by

$$\left(\frac{\partial}{\partial t_*} f\right)^*(t, \xi) := \frac{\partial f^*}{\partial t}(t, \xi) \quad \text{for a.e. } (t, \xi) \in Z_R^d, \tag{2.85}$$

with the domain

$$\mathcal{D}(\tfrac{\partial}{\partial t_*}; M) = \Big\{ f \in L^2(M) : \ \mathrm{L}_R^* f^*(\cdot, \xi) \in AC([0, \pi]), \ \mathrm{L}_R^* f^* \text{ satisfies} \tag{2.86}$$

$$\text{conditions (2.81) and (2.82) for } \mu\text{-a.e. } \xi \in \mathbb{S}^{d-1}, \ \frac{\partial}{\partial t_*} f \in L^2(M) \Big\}.$$

The definition (2.85) of the radial differential operator $\frac{\partial}{\partial t_*}$ can be described concretely by the following commutative diagram.

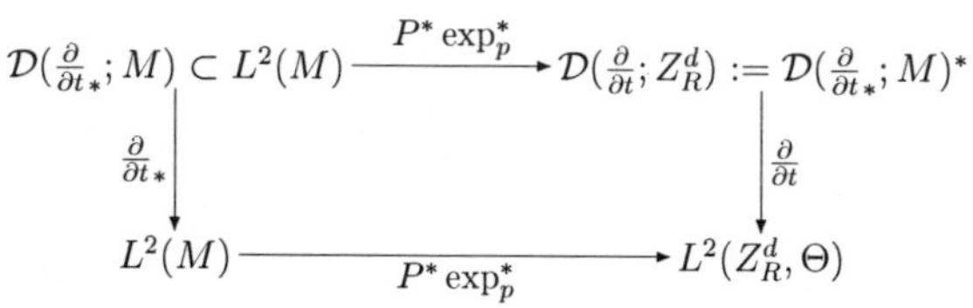

$$
\begin{array}{ccc}
\mathcal{D}(\tfrac{\partial}{\partial t_*}; M) \subset L^2(M) & \xrightarrow{\ P^* \exp_p^*\ } & \mathcal{D}(\tfrac{\partial}{\partial t}; Z_R^d) := \mathcal{D}(\tfrac{\partial}{\partial t_*}; M)^* \\[2pt]
{\scriptstyle \frac{\partial}{\partial t_*}}\Big\downarrow & & \Big\downarrow{\scriptstyle \frac{\partial}{\partial t}} \\[2pt]
L^2(M) & \xrightarrow[\ P^* \exp_p^*\ ]{} & L^2(Z_R^d, \Theta)
\end{array}
$$

Figure 7: Commutative diagram for the radial differential operator $\frac{\partial}{\partial t_*}$ on M.

The condition $\mathrm{L}_R^* f^*(\cdot, \xi) \in AC([0, \pi])$ for μ-a.e. $\xi \in \mathbb{S}^{d-1}$ in (2.86) is equivalent to the fact that the function f is absolutely continuous on μ-a.e. geodesic curve γ_ξ starting at $\gamma_\xi(0) = p$ in direction $\gamma_\xi'(0) = \xi$ and ending at the cut point $\gamma_\xi(R(\xi))$. Note that the exponential map in (2.58) was exactly defined by the geodesics γ_ξ. Further, the consistency conditions (2.81) and (2.82) ensure that for μ-a.e. $\xi_1, \xi_2 \in \mathfrak{S}_p$ the function values $f(\gamma_{\xi_1}(0))$ and $f(\gamma_{\xi_2}(0))$ coincide and that $f(\gamma_{\xi_1}(R(\xi_1)))$ coincides with $f(\gamma_{\xi_2}(R(\xi)))$ if $\gamma_{\xi_1}(R(\xi_1)) = \gamma_{\xi_2}(R(\xi_2))$ denotes the same point on the cut locus C_p of p.

In the definition (2.86) of the domain $\mathcal{D}(\tfrac{\partial}{\partial t_*}; M)$ is also implicitly stated that the functions $f^*(\cdot, \xi)$ are absolutely continuous on $[0, R(\xi)]$ for μ-a.e. $\xi \in \mathfrak{S}_p$ and that the derivative $\frac{\partial}{\partial t} f^*$ is an element of the Hilbert space $L^2(Z_R^d, \Theta)$. Hence, $\frac{\partial}{\partial t_*} f$ is a well defined function in $L^2(M)$ and describes precisely the derivative of f with respect to the geodesic distance t to the point p. Therefore, the denomination radial differential operator for the operator $\frac{\partial}{\partial t_*}$ is justified.

Moreover, since the continuously differentiable functions on M form a subspace of the domain $\mathcal{D}(\tfrac{\partial}{\partial t_*}; M)$, it follows from the Stone-Weierstrass Theorem B.1 that $\mathcal{D}(\tfrac{\partial}{\partial t_*}; M)$ is a dense subspace of $L^2(M)$. We can now define the following radial frequency variance for a function f on the manifold M.

Definition 2.39. We define the radial frequency variance $\text{var}_{F,p}^M(f)$ of a function $f \in \mathcal{D}(\frac{\partial}{\partial t_*}; M)$ as

$$\text{var}_{F,p}^M(f) := \left\| \frac{\partial}{\partial t_*} f \right\|_M^2. \tag{2.87}$$

Adopting the pull back operator $L_R^* P^* \exp_p^*$ to the derivative $\frac{\partial}{\partial t_*} f$, we get

$$L_R^* \left(\frac{\partial}{\partial t_*} f \right)^* = L_R^* \left(\frac{\partial}{\partial t} f^* \right) = \frac{\pi}{R} \frac{\partial}{\partial \tau} L_R^* f^*.$$

Hence, by Proposition 2.34, we can rewrite the latter definition as

$$\text{var}_{F,p}^M(f) = \left\| \frac{\pi}{R} \frac{\partial}{\partial \tau} L_R^* f^* \right\|_{W_{M,p}}^2.$$

Therefore, if we want to use $\text{var}_{F,p}^M(f)$ as a term for the frequency variance in Theorem 2.10, we have to choose the scaling function κ in (2.24) as $\kappa(\xi) = \frac{\pi}{R(\xi)}$. The respective Dunkl operator on $L^2(X^d, \tilde{W}_{M,p})$ can be introduced as follows.

Definition 2.40. Let $L^2(X^d, \tilde{W}_{M,p})$ be the extension of the Hilbert space $L^2(Z_\pi^d, W_{M,p})$ as in Definition 2.4. Then, we define the Dunkl operator $T_{M,p}^X$ on the Hilbert space $L^2(X^d, \tilde{W}_{M,p})$ as

$$T_{M,p}^X g := \frac{\pi}{R} \left(\frac{\partial g}{\partial \tau} + \frac{\tilde{W}_{M,p}'}{\tilde{W}_{M,p}} \frac{g - \check{g}}{2} \right), \tag{2.88}$$

with the domain

$$\mathcal{D}(T_{M,p}^X) := \Big\{ g \in L^2(X^d, \tilde{W}_{M,p}) : g(\cdot, \xi) \in AC_{2\pi} \quad \text{for } \mu\text{-a.e. } \xi \in \mathbb{S}^{d-1}, \tag{2.89}$$

$$\frac{\partial g}{\partial \tau}, \frac{W_{M,p}'}{W_{M,p}} \frac{g - \check{g}}{2} \in L^2(X^d, \tilde{W}_{M,p}) \Big\}.$$

The definition (2.88) of the Dunkl operator $T_{M,p}^X$ corresponds to the definition (2.16) of the Dunkl operator T^X with scaling function $\kappa = \frac{\pi}{R}$ and weight function $\tilde{W} = \tilde{W}_{M,p}$. The domain $\mathcal{D}(T_{M,p}^X)$ corresponds to the domain $\mathcal{D}(T^X)$ defined in (2.17). We get now, as a main result of this chapter, an uncertainty inequality for compact Riemannian manifolds.

Theorem 2.41.
Let M be a simply connected, compact Riemannian manifold without boundary and $p \in M$. If $f \in L^2(M) \cap \mathcal{D}(\frac{\partial}{\partial t_*}; M)$ such that $\|f\|_M = 1$, then, the following uncertainty principle holds:

$$\left(1 - \left(\int_{\mathfrak{S}_p} \int_0^{R(\xi)} \cos(\tfrac{\pi t}{R(\xi)}) |f^*(t,\xi)|^2 \Theta(t,\xi) dt d\mu(\xi) \right)^2 \right) \cdot \left\| \frac{\partial}{\partial t_*} f \right\|_M^2 \geq \tag{2.90}$$

$$\frac{1}{4} \left| \int_{\mathfrak{S}_p} \int_0^{R(\xi)} \left(\frac{\pi}{R(\xi)} \cos(\tfrac{\pi t}{R(\xi)}) + \sin(\tfrac{\pi t}{R(\xi)}) \frac{\Theta'(t,\xi)}{\Theta(t,\xi)} \right) \Theta(t,\xi) |f^*(t,\xi)|^2 dt d\mu(\xi) \right|^2.$$

Equality in (2.90) is attained if and only if f is constant.

Proof. If the function $f \in L^2(M) \cap \mathcal{D}(\frac{\partial}{\partial t_*}; M)$, then the pull back $L_R^* f^*$ lies in $L^2(Z_\pi^d, W_{M,p}) \cap \mathcal{D}(\frac{\partial}{\partial \tau}; Z_\pi^d)$ (see definition (2.23)). Further, the even extension $e(L_R^* f^*) \in L_e^2(X^d, \tilde{W}_{M,p})$ is an element of the domain $\mathcal{D}(T_{M,p}^X)$ of the Dunkl operator $T_{M,p}^X$. By Lemma 2.36, we know that the weight function $W_{M,p}$ satisfies Assumption 2.1. Now, using Theorem 2.10 together with the differential-difference operator $T_{M,p}^X$, yields the inequality

$$\left(1 - \left(\int_{\mathbb{S}^{d-1}} \int_0^\pi \cos\tau \, |L_R^* f^*(\tau, \xi)|^2 W_{M,p}(\tau, \xi) d\tau d\mu(\xi)\right)^2\right) \cdot \left\|\frac{\pi}{R(\xi)} \frac{\partial L_R^* f^*}{\partial \tau}\right\|_{W_{M,p}}^2 \ge$$
$$\frac{1}{4} \left|\int_{\mathbb{S}^{d-1}} \int_0^\pi \frac{\pi}{R(\xi)} \Big(\cos\tau \, W_{M,p}(\tau, \xi) + \sin\tau \, W'_{M,p}(\tau, \xi)\Big) |L_R^* f^*(\tau, \xi)|^2 dt d\mu(\xi)\right|^2.$$

Moreover, for the pull back operator L_R^*, we have $\tau = L_R^*(\frac{t\pi}{R})$, $\frac{\pi}{R}\frac{\partial}{\partial \tau} L_R^* f^* = L_R^*(\frac{\partial}{\partial t} f^*)$, $\frac{R}{\pi} W_{M,p} = L_R^*(\Theta)$ and $W'_{M,p} = L_R^*(\Theta')$. Hence, a coordinate transform with respect to the lipeomorphism L_R implies inequality (2.90). Finally, by Theorem 2.10, equality in (2.90) holds if and only if $f^*(\cdot, \xi) = C_\xi$ is constant for μ-a.e. $\xi \in \mathfrak{S}_p$. Since $f \in \mathcal{D}(\frac{\partial}{\partial t_*}; M)$ satisfies the consistency condition (2.81) for μ-a.e. $\xi \in \mathfrak{S}_p$, we have $C_\xi = C$ and f has to be constant μ_M-a.e. on M in order to obtain equality in (2.90). $\square$

Similar to the Breitenberger uncertainty principle (1.18) and to the uncertainty principle (1.38) for weighted L^2-spaces on the interval, we can introduce a generalized mean value $\varepsilon_p(f)$ for a function $f \in L^2(M)$ by

$$\varepsilon_p(f) := \langle e^{i\tau} e(L_R^* f^*), e(L_R^* f^*)\rangle_{\tilde{W}_{M,p}} \tag{2.91}$$
$$= \int_{\mathfrak{S}_p} \int_0^{R(\xi)} \cos(\tfrac{\pi t}{R(\xi)}) |f^*(t, \xi)|^2 \Theta(t, \xi) dt d\mu(\xi).$$

Moreover, we denote the integral term on the right hand side of (2.90) as

$$\rho_p(f) := \langle -(i T_{M,p}^X e^{i\tau}) e(L_R^* f^*), e(L_R^* f^*)\rangle_{\tilde{W}_{M,p}} \tag{2.92}$$
$$= \int_{\mathfrak{S}_p} \int_0^{R(\xi)} \left(\tfrac{\pi}{R(\xi)} \cos(\tfrac{\pi t}{R(\xi)}) + \sin(\tfrac{\pi t}{R(\xi)}) \tfrac{\Theta'(t,\xi)}{\Theta(t,\xi)}\right) |f^*(t, \xi)|^2 \Theta(t, \xi) dt d\mu(\xi).$$

Definition 2.42. Let $f \in L^2(M)$. If $\rho_p(f) \ne 0$, we define

$$\text{var}_{S,p}^M(f) := d^2 \frac{1 - \varepsilon_p(f)^2}{\rho_p(f)^2}. \tag{2.93}$$

The value $\text{var}_{S,p}^M(f)$ is called the position variance of the function f at the point $p \in M$.

Corollary 2.43.
Let $p \in M$ and $f \in L^2(M) \cap \mathcal{D}(\frac{\partial}{\partial t_*}; M)$ with $\|f\|_M = 1$ and $\rho_p(f) \ne 0$. Then,

$$\text{var}_{S,p}^M(f) \cdot \text{var}_{F,p}^M(f) > \frac{d^2}{4}. \tag{2.94}$$

Proof. Evidently, (2.94) follows from (2.90). The only thing that remains to check is the strict inequality in (2.94). The only functions for which equality can be obtained in (2.90) are the constant functions. If $f = C$ is constant on M, integration by parts with respect to the variable τ yields

$$\rho_p(f) = \int_{\mathfrak{S}_p} \int_0^{R(\xi)} \left(\tfrac{\pi}{R(\xi)} \cos(\tfrac{\pi t}{R(\xi)}) \Theta(t,\xi) + \sin(\tfrac{\pi t}{R(\xi)}) \Theta'(t,\xi) \right) dt d\mu(\xi)$$
$$= \int_{\mathbb{S}^{d-1}} \int_0^{\pi} (\cos \tau \, W_{M,p}(\tau) + \sin \tau \, W'_{M,p}(\tau)) d\tau d\mu(\xi) = 0.$$

Hence, there exists no function f with $f \in L^2(M) \cap \mathcal{D}(\frac{\partial}{\partial t_*}; M)$, $\|f\|_M = 1$ and $\rho_p(f) \neq 0$ for which equality can be attained in (2.94). $\square$

Remark 2.44. Although inequality (2.94) is strict, we will show in Proposition 2.58 that the constant $\frac{d^2}{4}$ on the right hand side of (2.94) is optimal.

In contrast to Theorem 2.10 where we considered relatively general weight functions W on Z_π^d, the weight function $W_{M,p}$ and its counterpart Θ in Theorem 2.41 play a more substantial role. These weight functions are linked to the exponential map $\exp_p$ on the tangent space $T_p M$ and implicitly contain information on the curvature of the Riemannian manifold M at the point p. We will see in Section 2.7 how this information can be used to compute lower estimates of inequality (2.90).

Remark 2.45. Using the generalized mean value $\varepsilon_p(f)$, we can search for a point $p_f \in M$ that can be interpreted as the expectation value of the density $f \in L^2(M)$, $\|f\|_M = 1$. Namely, we consider the value $\varepsilon_p(f)$ as a measure on how well the function f is localized at the point $p \in M$. Since $\|f\|_M = 1$, the closer $\varepsilon_p(f)$ approaches the value 1, the more the L^2-mass of f is concentrated at p. The point at which f is localized best is then defined as the point p_f where $\varepsilon_p(f)$ gets maximal, i.e.,

$$p_f = \arg\sup_{p \in M} \varepsilon_p(f). \tag{2.95}$$

If p_f is uniquely determined, we call it the expectation value of f.

2.2.3. The Dunkl and the radial Laplace-Beltrami operator

The Dunkl operator $T_{M,p}^X$ is closely related to the Laplace-Beltrami operator Δ_M of the Riemannian manifold M (see Section A.7 for a short introduction and the definition). For a radial function F centered at the point $p \in M$, i.e., the pull back F^* depends solely on the radial distance t, the Laplace-Beltrami operator Δ_M reads in a small neighborhood around p as (cf. [3, Proposition G.V.3])

$$(\Delta_M F)^*(t,\xi) = \frac{d^2}{dt^2} F^*(t) + \frac{\Theta'(t,\xi)}{\Theta(t,\xi)} \frac{d}{dt} F^*(t).$$

This second-order differential operator can be extended to the whole manifold M and used globally for functions f on M. In geodesic polar coordinates $(t, \xi) \in Z_R^d$, we define the operator $\Delta_{p,t}$ as

$$(\Delta_{p,t} f)^*(t, \xi) := \frac{\partial^2}{\partial t^2} f^*(t, \xi) + \frac{\Theta'(t, \xi)}{\Theta(t, \xi)} \frac{\partial}{\partial t} f^*(t, \xi). \tag{2.96}$$

For radial functions centered at the point p, the operator $\Delta_{p,t}$ corresponds locally with the Laplace-Beltrami operator Δ_M. Therefore, the operator $\Delta_{p,t}$ is referred to as radial Laplace-Beltrami operator. Adopting the lipeomorphism L_R, we get for $(\tau, \xi) \in Z_\pi^d$ the formula

$$L_R^*(\Delta_{p,t} f)^*(\tau, \xi) = \frac{\pi^2}{R(\xi)^2} \left(\frac{\partial^2}{\partial \tau^2} L_R^* f^*(\tau, \xi) + \frac{W'_{M,p}(\tau, \xi)}{W_{M,p}(\tau, \xi)} \frac{\partial}{\partial \tau} L_R^* f^*(\tau, \xi) \right).$$

As a domain of the radial Laplacian, we consider the set

$$\mathcal{D}(\Delta_{p,t}) := \left\{ f \in C^2(M) : \frac{\partial}{\partial \tau} L_R^* f^*(0, \xi) = \frac{\partial}{\partial \tau} L_R^* f^*(\pi, \xi) = 0, \ \xi \in \mathbb{S}^{d-1} \right\}. \tag{2.97}$$

So, if $f \in \mathcal{D}(\Delta_{p,t})$, then $iT_{M,p}^X(e(L_R^* f^*)) = i\frac{\pi}{R} \frac{\partial e(L_R^* f^*)}{\partial \tau} \in \mathcal{D}(T_{M,p}^X)$, and we get the following relation between the radial Laplace-Beltrami-operator $\Delta_{p,t}$ and the Dunkl operator $iT_{M,p}^X$ visualized in Figure 8:

$$e\left(L_R^*(-\Delta_{p,t} f)^*\right) = (iT_{M,p}^X)^2 e(L_R^* f^*). \tag{2.98}$$

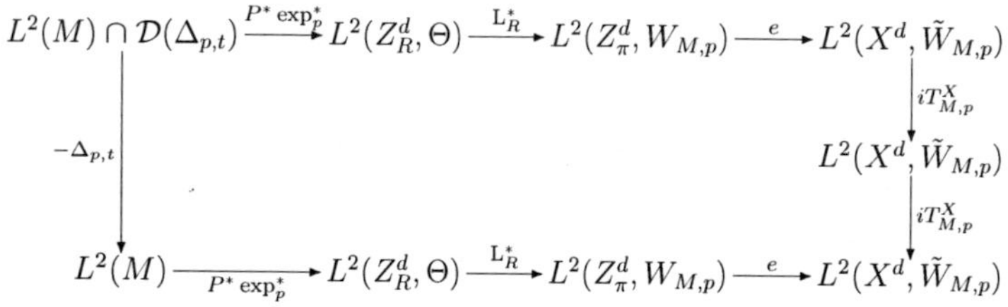

Figure 8: Commutative diagram for the decomposition of $-\Delta_{p,t}$.

In this way, the differential-difference operator $iT_{M,p}^X$ can be seen as a generalized symmetric root of the operator $-\Delta_{p,t}$. In particular, this relation gives a new view on the frequency variance $\mathrm{var}_{F,p}^M(f)$ of a function $f \in \mathcal{D}(\Delta_{p,t})$ defined in 2.87. Namely, we get

$$\begin{aligned}
\mathrm{var}_{F,p}^M(f) &= \left\| \frac{\partial}{\partial t_*} f \right\|_M^2 = \left\| iT_{M,p}^X e(L_R^* f^*) \right\|_{\tilde{W}_{M,p}}^2 \\
&= \left\langle (iT_{M,p}^X)^2 e(L_R^* f^*), e(L_R^* f^*) \right\rangle_{\tilde{W}_{M,p}} = \langle -\Delta_{p,t} f, f \rangle_M.
\end{aligned} \tag{2.99}$$

Formula (2.99) asserts that the frequency variance in the uncertainty inequality (2.94) is completely determined by the radial derivative of the function f. Many authors (see, for instance, [27] or [56]) prefer to use the full Laplace-Beltrami operator Δ_M for the frequency variance, i.e. $\mathrm{var}_F^M(f) := \langle -\Delta_M f, f \rangle_M$, instead of the radial approach (2.99). However, since $\mathrm{var}_{F,p}^M(f) \leq \langle -\Delta_M f, f \rangle_M$ for functions f that are locally supported at $p \in M$, we get sharper inequalities in (2.90) and (2.94) if we use the radial Laplace-Beltrami operator.

2.3. Uncertainty principles on compact star-shaped domains

In this section, we will give an alternative uncertainty principle in the case that the underlying domain of the L^2-space is not a whole Riemannian manifold M but a compact subset $\Omega \subset M$. We will only consider functions that satisfy a zero boundary condition at the boundary $\partial\Omega$ of Ω. The goal is to establish an uncertainty principle for locally supported functions with a position variance that is easier to handle than the position variance (2.93) in the last section. For the proof of this uncertainty principle, we want to adopt Theorem 2.14. Therefore, we have to show as in the last section that the Hilbert space $L^2(\Omega)$ is isometrically isomorphic to a weighted space $L^2(Z_\pi^d, W_{\Omega,p})$. This is possible if we assume that the compact set Ω is star-shaped with respect to a point $p \in M$ and that its boundary $\partial\Omega$ satisfies a Lipschitz condition.

Definition 2.46. We call a compact subset Ω of a Riemannian manifold M star-shaped with respect to an interior point $p \in \Omega$ if:

(i) For every point $q \in \Omega$ there exists a minimizing geodesic γ_ξ with $\gamma_\xi(0) = p$ and $\gamma_\xi(t_q) = q$ such that $\gamma_\xi(t) \in \Omega$ for all $t \in [0, t_q]$.

(ii) If $q \in \partial\Omega$ is an element of the boundary $\partial\Omega$ of Ω, then $\gamma_\xi(t) \notin \partial\Omega$ lies in the interior of Ω for all $t \in [0, t_q)$.

By $Q(\xi)$, we denote the length $d(p, q) > 0$ of the geodesic γ_ξ connecting the center point p with a boundary point $q \in \partial\Omega$ in direction $\xi \in \mathfrak{S}_p$. From now on, we will assume that Ω is a compact star-shaped subset of a (not necessarily compact) Riemannian manifold M and that the distance function Q is Lipschitz continuous on $\mathfrak{S}_p \subset T_pM$.

By $L^2(\Omega)$, we denote the Hilbert space of square integrable functions on Ω with scalar product

$$\langle f, g \rangle_\Omega := \int_\Omega f(q)\overline{g(q)} d\mu_M(q) \tag{2.100}$$

and norm $\|f\|_\Omega^2 := \langle f, f \rangle_\Omega$. To show that $L^2(\Omega)$ is isometrically isomorphic to a weighted space $L^2(Z_\pi^d, W_{\Omega,p})$, we use similar as in Section 2.2 three coordinate transforms: the exponential map $\exp_p$, the polar transform P and a Lipschitz continuous mapping L_Ω.

Let Z_Q^d be defined as in (2.67) with the distance function Q instead of the distance R and the boundaries $\partial_L Z_Q^d$ and $\partial_R Z_Q^d$. Since Ω is a star-shaped and compact set, the composition $\exp_p \mathrm{P}$ is a well defined function on Z_Q^d that maps the set $Z_Q^d \setminus \{\partial_L Z_Q^d, \partial_R Z_Q^d\}$ diffeomorphically onto the domain $\Omega \setminus \{p, \partial\Omega\}$. Further, the left hand boundary $\partial_L Z_Q^d$ is mapped onto $\{p\}$ and the right hand boundary $\partial_R Z_Q^d$ onto the boundary $\partial\Omega$ of Ω. So the points $(t, \xi) \in Z_Q^d$ form a coordinate system for the domain Ω, referred to as geodesic polar coordinates on Ω. Moreover, for an integrable function f on Ω, we get from (A.30) and the polar transform P the integral formula

$$\int_\Omega f(q) d\mu_M(q) = \int_{\mathfrak{S}_p} \int_0^{Q(\xi)} f^*(t, \xi)\Theta(t, \xi) dt d\mu(\xi), \qquad (2.101)$$

where the Jacobi determinant Θ is given as in (2.73).

Let $\mathrm{L}_Q : Z_\pi^d \to Z_Q^d$ be the coordinate transformation as defined in (2.76) with the distance function Q instead of R. Then, since the distance function Q is assumed to be Lipschitz continuous on $\mathfrak{S}_p$, the coordinate transform L_Q is a lipeomorphism that maps Z_π^d onto Z_Q^d. Further, L_Q induces a change of variables that leads, as in (2.77), to the integral formula

$$\int_{\mathfrak{S}_p} \int_0^{Q(\xi)} f^*(t, \xi)\Theta(t, \xi) dt d\mu(\xi) = \int_{\mathbb{S}^{d-1}} \int_0^\pi f^*(\tfrac{Q(\xi)}{\pi}\tau, \xi)\tfrac{\pi}{Q(\xi)}\Theta(\tfrac{Q(\xi)}{\pi}\tau, \xi) d\tau d\mu(\xi). \quad (2.102)$$

Hereby, the Jacobian determinant of L_Q equals $\frac{\pi}{Q}$ almost everywhere on Z_π^d.

For a function f^* on Z_Q^d, we define the pull back $\mathrm{L}_Q^* f^*$ by the lipeomorphism L_Q as in (2.78) by

$$\mathrm{L}_Q^* f^*(\tau, \xi) := f^*(\tfrac{Q(\xi)}{\pi}\tau, \xi), \quad (\tau, \xi) \in [0, \pi] \times \mathbb{S}^{d-1}, \qquad (2.103)$$

and introduce the weight function $W_{\Omega, p}$ on Z_π^d as

$$W_{\Omega, p}(\tau, \xi) := \tfrac{\pi}{Q(\xi)}\Theta(\tfrac{Q(\xi)}{\pi}\tau, \xi), \quad (\tau, \xi) \in [0, \pi] \times \mathbb{S}^{d-1}. \qquad (2.104)$$

Then, analogously to Proposition 2.34, we get the following result.

Proposition 2.47.
The Hilbert space $L^2(\Omega)$ is isometrically isomorphic to the space $L^2(Z_\pi^d, W_{\Omega, p})$. The isomorphism is given by the pull back operator $\mathrm{L}_Q^* \mathrm{P}^* \exp_p^*$.

Proof. We collect the coordinate changes given by the exponential map $\exp_p$, the polar transform P and the lipeomorphism L_Q. For f, g in $L^2(\Omega)$, we get

$$\langle f, g \rangle_\Omega \overset{(2.100)}{=} \int_\Omega f(q)\overline{g(q)} d\mu_M(q)$$

$$\overset{(2.101)}{=} \int_{\mathfrak{S}_p} \int_0^{Q(\xi)} f^*(t, \xi)\overline{g^*(t, \xi)}\Theta(t, \xi) dt d\mu(\xi)$$

$$\overset{(2.102)}{=} \int_{\mathbb{S}^{d-1}} \int_0^\pi f^*\left(\frac{Q(\xi)}{\pi}\tau,\xi\right)\overline{g^*\left(\frac{Q(\xi)}{\pi}\tau,\xi\right)}\frac{\pi}{Q(\xi)}\Theta\left(\frac{Q(\xi)}{\pi}\tau,\xi\right)d\tau d\mu(\xi)$$

$$\overset{(2.103)}{\underset{(2.104)}{=}} \int_{\mathbb{S}^{d-1}} \int_0^\pi \mathrm{L}_Q^* f^*(\tau,\xi)\overline{\mathrm{L}_Q^* g^*(\tau,\xi)}W_{\Omega,p}(\tau,\xi)d\tau d\mu(\xi)$$

$$\overset{(2.7)}{=} \langle \mathrm{L}_Q^* f^*, \mathrm{L}_Q^* g^*\rangle_{W_{\Omega,p}}.$$

Hence, the pull back operator $\mathrm{L}_Q^* \mathrm{P}^* \exp_p^*$ is an isometric homomorphism from $L^2(\Omega)$ to $L^2(Z_\pi^d, W_{\Omega,p})$. Since the point set $\{p\}$ and the boundary $\partial\Omega$ are sets of μ_M-measure zero in Ω and $\exp_p \mathrm{PL}_Q$ defines a lipeomorphism from $Z_\pi^d \setminus \{\partial_L Z_\pi^d, \partial_R Z_\pi^d\}$ onto $\Omega \setminus \{p, \partial\Omega\}$, the operator $\mathrm{L}_Q^* \mathrm{P}^* \exp_p^*$ from $L^2(\Omega)$ to the space $L^2(Z_\pi^d, W_{\Omega,p})$ is also surjective and, thus, an isomorphism. $\qquad\square$

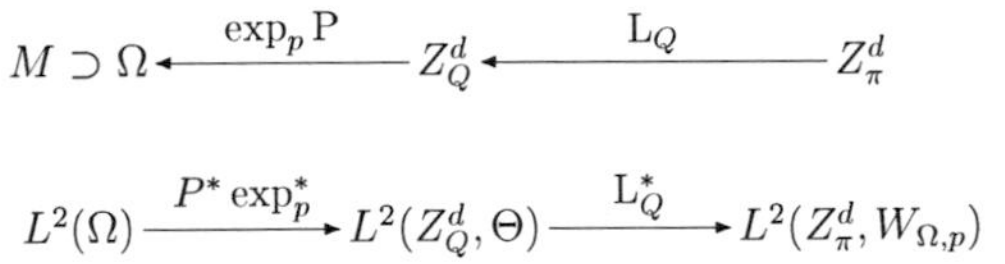

$$M \supset \Omega \xleftarrow{\quad \exp_p \mathrm{P} \quad} Z_Q^d \xleftarrow{\quad \mathrm{L}_Q \quad} Z_\pi^d$$

$$L^2(\Omega) \xrightarrow{\quad \mathrm{P}^* \exp_p^* \quad} L^2(Z_Q^d, \Theta) \xrightarrow{\quad \mathrm{L}_Q^* \quad} L^2(Z_\pi^d, W_{\Omega,p})$$

Figure 9: The mappings $\exp_p \mathrm{P}$, L_Q and the respective pull backs.

As in Lemma 2.36, we can now show that the weight function $W_{\Omega,p}$ satisfies the required Assumption 2.1.

Lemma 2.48.
The weight function $W_{\Omega,p}$ on Z_π^d satisfies Assumption 2.1.

Proof. To prove Lemma 2.48, we just have to follow the lines of the proof of Lemma 2.36 and replace the cut locus distance R by the distance function Q. $\qquad\square$

Now, in order to use Theorem 2.14, we have to guarantee that for $f \in L^2(\Omega)$ the pull back $\mathrm{L}_Q^* f^*$ lies in $\mathcal{D}_0(\frac{\partial}{\partial t}; Z_\pi^d)$, as defined in (2.28). In particular, the function $\mathrm{L}_Q^* f^*$ has to satisfy a zero boundary condition on $\partial_R Z_\pi^d$. Moreover, we want to make sure that $\mathrm{L}_Q^* f^*$ fulfills the consistency condition (2.81), i.e. that $\mathrm{L}_Q^* f^*(0,\xi_1) = \mathrm{L}_Q^* f^*(0,\xi_2)$ holds for μ-a.e. $\xi_1, \xi_2 \in \mathbb{S}^{d-1}$. Altogether, we define in the style of (2.86) the domain of the radial differential operator $\frac{\partial}{\partial t_*}$ on $\Omega \subset M$ as

$$\mathcal{D}_0(\tfrac{\partial}{\partial t_*}; \Omega) := \Big\{ f \in L^2(\Omega) : \mathrm{L}_Q^* f^*(\cdot,\xi) \in AC([0,\pi]), \ \mathrm{L}_Q^* f^* \text{ satisfies (2.81) and}$$

$$\mathrm{L}_Q^* f^*(\pi,\xi) = 0 \ \mu\text{-a.e.}, \ \frac{\partial}{\partial\tau}\mathrm{L}_Q^* f^*, \ \tau\frac{W_{\Omega,p}'}{W_{\Omega,p}}\mathrm{L}_Q^* f^* \in L^2(Z_\pi^d, W_{\Omega,p}) \Big\}. \quad (2.105)$$

By Proposition 2.47, the condition $\frac{\partial}{\partial\tau}\mathrm{L}_Q^* f^* \in L^2(Z_\pi^d, W_{\Omega,p})$ is equivalent to the property $\frac{\partial}{\partial t_*}f \in L^2(\Omega)$. As a position and frequency variance of a function f on Ω with respect to a point p, we will use the following expressions:

Definition 2.49. We define the position variance $\text{var}^\Omega_{S,p}(f)$ and the radial frequency variance $\text{var}^\Omega_{F,p}(f)$ of a function $f \in \mathcal{D}_0(\frac{\partial}{\partial t_*}; \Omega)$ as

$$\text{var}^\Omega_{S,p}(f) := \|t_* f\|^2_\Omega, \tag{2.106}$$

$$\text{var}^\Omega_{F,p}(f) := \left\| \frac{\partial}{\partial t_*} f \right\|^2_\Omega. \tag{2.107}$$

Due to Proposition 2.47, the radial frequency variance can be rewritten as

$$\text{var}^\Omega_{F,p}(f) = \left\| \frac{\pi}{Q} \frac{\partial}{\partial \tau} \text{L}^*_Q f^* \right\|^2_{W_{\Omega,p}}.$$

Therefore, in order to use the expression $\text{var}^\Omega_{F,p}(f)$ in Theorem 2.14 we have to choose the scaling function κ as $\kappa(\xi) = \frac{\pi}{Q(\xi)}$. Related to this scaling function is the following Dunkl operator:

Definition 2.50. Let $L^2(X^d, \tilde{W}_{\Omega,p})$ be the extension of the Hilbert space $L^2(Z^d_\pi, W_{\Omega,p})$ as in Definition 2.4. Then, we define the Dunkl operator $T^X_{\Omega,p}$ on $L^2(X^d, \tilde{W}_{\Omega,p})$ as

$$T^X_{\Omega,p} g := \frac{\pi}{Q} \left(\frac{\partial g}{\partial \tau} + \frac{\tilde{W}'_{\Omega,p}}{\tilde{W}_{\Omega,p}} \frac{g - \breve{g}}{2} \right), \tag{2.108}$$

with the domain

$$\mathcal{D}_0(T^X_{\Omega,p}) := \Big\{ g \in L^2(X^d, \tilde{W}_{\Omega,p}) : \ g(\cdot, \xi) \in AC_{2\pi}, \ g(\pi, \xi) = 0 \text{ for } \mu\text{-a.e. } \xi \in \mathbb{S}^{d-1},$$

$$\frac{\partial g}{\partial \tau}, \frac{W'_{\Omega,p}}{W_{\Omega,p}} \frac{g - \breve{g}}{2}, \ \tau \frac{W'_{\Omega,p}}{W_{\Omega,p}} g \in L^2(X^d, \tilde{W}_{\Omega,p}) \Big\}. \tag{2.109}$$

The definition (2.108) of the Dunkl operator $T^X_{\Omega,p}$ corresponds to the definition (2.16) of T^X with the scaling function $\kappa(\xi) = \frac{\pi}{Q(\xi)}$ and the weight function $\tilde{W} = \tilde{W}_{\Omega,p}$. The domain $\mathcal{D}_0(T^X_{\Omega,p})$ corresponds to the restricted domain $\mathcal{D}_0(T^X)$ defined in (2.25). By Lemma 2.8, we know that $iT^X_{\Omega,p}$ is symmetric on $L^2(X^d, \tilde{W}_{\Omega,p})$. Hence, we get the following uncertainty principle for compact star-shaped domains:

Theorem 2.51.
Let M be a Riemannian manifold and $\Omega \subset M$ be a compact star-shaped domain with interior point p and Lipschitz continuous boundary $\partial\Omega$. Let $f \in L^2(\Omega) \cap \mathcal{D}_0(\frac{\partial}{\partial t_*}; \Omega)$ such that $\|f\|_\Omega = 1$. Then, the following uncertainty principle holds:

$$\|t_* f\|^2_\Omega \cdot \left\| \frac{\partial}{\partial t_*} f \right\|^2_\Omega > \frac{1}{4} \left| 1 + \int_{\mathfrak{S}_p} \int_0^{Q(\xi)} t\Theta'(t,\xi)) |f^*(t,\xi)|^2 dt d\mu(\xi) \right|^2. \tag{2.110}$$

Proof. If $f \in L^2(\Omega) \cap \mathcal{D}_0(\frac{\partial}{\partial t_*}; \Omega)$, then, by Proposition 2.47, $\text{L}^*_Q f^* \in L^2(Z^d_\pi, W_{\Omega,p}) \cap \mathcal{D}_0(\frac{\partial}{\partial \tau}; Z^d_\pi)$ and the even extension $e(\text{L}^*_Q f^*) \in L^2_e(X^d, \tilde{W}_{\Omega,p})$ is an element of the domain

$\mathcal{D}_0(T^X_{\Omega,p})$ of the Dunkl operator $T^X_{\Omega,p}$. Moreover, we know from Lemma 2.48 that the weight function $W_{\Omega,p}$ satisfies Assumption 2.1. Then, if we consider the symmetric operators A and B on $L^2(X^d, \tilde{W}_{\Omega,p})$ defined by $Ag = \frac{Q(\xi)}{\pi}\tau g$ and $Bg = iT^X_{\Omega,p}g$, we get in Theorem 2.14 the inequality

$$\left\|\frac{Q(\xi)}{\pi}\tau L^*_Q f^*\right\|^2_{W_{\Omega,p}} \cdot \left\|\frac{\pi}{Q(\xi)}\frac{\partial L^*_Q f^*}{\partial\tau}\right\|^2_{W_{\Omega,p}} > $$
$$\frac{1}{4}\left|\int_{\mathbb{S}^{d-1}}\int_0^\pi (W_{\Omega,p}(\tau,\xi) + \tau W'_{\Omega,p}(\tau,\xi))|L^*_Q f^*(\tau,\xi)|^2 d\tau d\mu(\xi)\right|^2.$$

Now, by the coordinate transformation L_Q, we have $\tau = L^*_Q(\frac{t\pi}{Q})$, $\frac{\pi}{Q}\frac{\partial}{\partial\tau}L^*_Q f^* = L^*_Q(\frac{\partial}{\partial t}f^*)$ and $W'_{\Omega,p} = L^*_Q(\Theta')$. This implies inequality (2.110). $\qquad\square$

Example 2.52. As an example of a compact star-shaped domain we consider the unit ball B^d in $\mathbb{R}^d$ centered at $p = 0$. In this case, the distance function Q is given by $Q(\xi) = 1$ for all $\xi \in \mathfrak{S}_p$ and the weight function Θ on Z^d_Q is given by $\Theta(t,\xi) = t^{d-1}$. Hence $\Theta'(t,\xi) = (d-1)t^{d-2}$ and Theorem 2.51 implies the inequality

$$\|t_* f\|^2_{B^d} \cdot \left\|\frac{\partial}{\partial t_*}f\right\|^2_{B^d} > \frac{d^2}{4} \tag{2.111}$$

for all functions $f \in L^2(B^d) \cap \mathcal{D}_0(\frac{\partial}{\partial t_*}; B^d)$ normalized such that $\|f\|_{B^d} = 1$.

2.4. Uncertainty principles on manifolds diffeomorphic to $\mathbb{R}^d$

We leave now the compact settings and consider Riemannian manifolds E that are diffeomorphic to the Euclidean space $\mathbb{R}^d$. In this particular case, the exponential map $\exp_p$ defines a diffeomorphism from $T_p E$ onto E. Thus, by (A.30) and the polar transform P, we get for an integrable function f on E in geodesic polar coordinates $(t,\xi) \in [0,\infty) \times \mathfrak{S}_p$ the formula

$$\int_E f(q)d\mu_E(q) = \int_{\mathfrak{S}_p}\int_0^\infty f^*(t,\xi)\Theta(t,\xi)dt d\mu(\xi), \tag{2.112}$$

where the weight function Θ is given as in (2.73). To keep the notation simple, we identify the tangent space $T_p E$ with the Euclidean space $\mathbb{R}^d$ and use the symbol Z^d_∞ instead of $[0,\infty) \times \mathfrak{S}_p$. However, to indicate that we are working in the tangent space $T_p E$, we will still use the symbol $\mathfrak{S}_p$ for the unit sphere.

The Hilbert space $L^2(E)$ with inner product

$$\langle f,g\rangle_E := \int_E f(q)\overline{g(q)}d\mu_E(q) \tag{2.113}$$

is isometrically isomorphic to the Hilbert space $L^2(Z^d_\infty, \Theta)$ with scalar product

$$\langle f^*, g^*\rangle_\Theta := \int_{\mathfrak{S}_p}\int_0^\infty f^*(t,\xi)\overline{g^*(t,\xi)}\Theta(t,\xi)dt d\mu(\xi). \tag{2.114}$$

Now, we can derive an uncertainty principle on the non-compact Riemannian manifold E by using the uncertainty principle on the tube Z_∞^d developed in Section 2.1.3. In particular, we can adopt Theorem 2.21. First, we show that the weight function Θ is admissible.

Lemma 2.53.
Let E be diffeomorphic to the Euclidean space $\mathbb{R}^d$. Then the weight function Θ satisfies Assumption 2.15.

Proof. Since E is diffeomorphic to $\mathbb{R}^d$, the Jacobi determinant $\theta(\xi) = \det((d\exp_p)_\xi)$ is continuously differentiable and strictly positive on the whole tangent space $T_p E$. Hence $\Theta(t,\xi) = t^{d-1}\theta(t\xi)$ is continuously differentiable on Z_∞^d and vanishes only at the boundary $\partial_L Z_\infty^d$ of Z_∞^d. Hence, the conditions (2.32) and (2.33) are satisfied. Further, since

$$t\frac{\Theta'(t,\xi)}{\Theta(t,\xi)} = (d-1) + t\frac{\theta'(t\xi)}{\theta(t\xi)}$$

is bounded for every compact subset of Z_∞^d, also (2.34) is satisfied. □

From Theorem 2.21, we can now derive the following uncertainty principle for Riemannian manifolds diffeomorphic to the Euclidean space.

Theorem 2.54.
Let E be a Riemannian manifold diffeomorphic to $\mathbb{R}^d$ and $p \in E$. Let $f \in L^2(E)$, $\|f\|_E = 1$, such that $f^* \in \mathcal{D}(\frac{\partial}{\partial t}, t, t\frac{\partial}{\partial t}; Z_\infty^d)$ (see (2.46)) and such that the consistency condition $f^*(0,\xi_1) = f^*(0,\xi_2)$ is satisfied for μ-a.e. $\xi_1, \xi_2 \in \mathfrak{S}_p$. Then, the following uncertainty principle holds:

$$\|t_* f\|_E^2 \cdot \left\|\frac{\partial}{\partial t_*} f\right\|_E^2 \geq \frac{1}{4}\left|1 + \int_{\mathfrak{S}_p}\int_0^\infty t\Theta'(t,\xi)|f^*(t,\xi)|^2 dt d\mu(\xi)\right|^2. \tag{2.115}$$

Equality holds if and only if $f^*(t,\xi) = Ce^{-\lambda t^2}$, with a complex scalar C and a real constant $\lambda \in \mathbb{R}$ such that f satisfies the requirements of the theorem.

Proof. By Lemma 2.53, the weight function Θ satisfies Assumption 2.15 and $\|f^*\|_\Theta = \|f\|_E = 1$. Thus, the statement follows from Theorem 2.21. Hereby, the consistency condition $f^*(0,\xi_1) = f^*(0,\xi_2)$ for μ-a.e. $\xi_1, \xi_2 \in \mathfrak{S}_p$ makes sure that the complex constant C in the optimal function $f^*(t,\xi) = Ce^{-\lambda t^2}$ does not depend on the variable $\xi \in \mathfrak{S}_p$. □

Definition 2.55. In the non-compact case, we define the position and frequency variance of a function f at the point $p \in E$ as

$$\mathrm{var}_{S,p}^E(f) := \|t_* f\|_E^2 = \int_{\mathfrak{S}_p}\int_0^\infty t^2 |f^*(t,\xi)|^2 \Theta(t,\xi) dt d\mu(\xi), \tag{2.116}$$

$$\mathrm{var}_{F,p}^E(f) := \left\|\frac{\partial}{\partial t_*} f\right\|_E^2 = \int_{\mathfrak{S}_p}\int_0^\infty \left|\frac{\partial f^*}{\partial t}(t,\xi)\right|^2 \Theta(t,\xi) dt d\mu(\xi). \tag{2.117}$$

In view of the latter definition, the following uncertainty inequality holds for all functions satisfying the requirements of Theorem 2.54.

$$\text{var}_{S,p}^E(f) \cdot \text{var}_{F,p}^E(f) \geq \frac{1}{4}\left|1 + \int_{\mathfrak{S}_p}\int_0^\infty t\Theta'(t,\xi)|f^*(t,\xi)|^2 dt d\mu(\xi)\right|^2. \tag{2.118}$$

Example 2.56. If $E = \mathbb{R}^d$, $p \in \mathbb{R}^d$, then $\Theta(t,\xi) = t^{d-1}$, $\Theta'(t,\xi) = (d-1)t^{d-2}$ and Theorem 2.54 implies the uncertainty principle

$$\|t_* f\|_{\mathbb{R}^d}^2 \cdot \left\|\frac{\partial}{\partial t_*} f\right\|_{\mathbb{R}^d}^2 \geq \frac{d^2}{4}. \tag{2.119}$$

Equality in (2.119) is attained if and only if $f(q) = Ce^{-\lambda|q|^2}$ for a constant $\lambda > 0$ and a complex constant C. This inequality is the d-dimensional analog of the one-dimensional Heisenberg-Pauli-Weyl inequality (1.9).

Similar as in the case of a compact manifold M, there exists a relation between the Dunkl operator T^Y given on $L^2(Y^d, \tilde{\Theta})$ by

$$T^Y g = \frac{\partial g}{\partial t} + \frac{\tilde{\Theta}'}{\tilde{\Theta}}\frac{g - \check{g}}{2}$$

and the radial part of the Laplace-Beltrami operator Δ_E. As in (2.96), we introduce the radial Laplace-Beltrami operator $\Delta_{p,t}$ in geodesic polar coordinates (t,ξ) as

$$(\Delta_{p,t}f)^*(t,\xi) := \frac{\partial^2}{\partial t^2}f^*(t,\xi) + \frac{\Theta'(t,\xi)}{\Theta(t,\xi)}\frac{\partial}{\partial t}f^*(t,\xi), \quad (t,\xi) \in Z_\infty^d, \tag{2.120}$$

on the domain

$$\mathcal{D}(\Delta_{p,t}) := \left\{f \in C^2(E) \cap L^2(E) : \frac{\partial}{\partial t}f^*(0,\xi) = 0, \ \xi \in \mathfrak{S}_p\right\}. \tag{2.121}$$

Now, for $f \in \mathcal{D}(\Delta_{p,t})$, we can use the even extension $e(f^*) \in L_e^2(Y^d, \tilde{\Theta})$ to get the following decomposition of the radial Laplace-Beltrami operator $\Delta_{p,t}$, also shown in the commutative diagram in Figure 10.

$$e((-\Delta_{p,t}f)^*) = (iT^Y)^2 e(f^*). \tag{2.122}$$

Hence, as in the compact setting, the operator iT^Y can be seen as a generalized symmetric root of $-\Delta_{p,t}$. Further, since

$$\text{var}_{S,p}^E(f) = \left\|\frac{\partial}{\partial t_*}f\right\|_E^2 = \left\|iT^Y e(f^*)\right\|_{\tilde{\Theta}}^2 = \langle-\Delta_{p,t}f, f\rangle_E, \tag{2.123}$$

we get a second representation of the frequency variance $\text{var}_{S,p}^E(f)$.

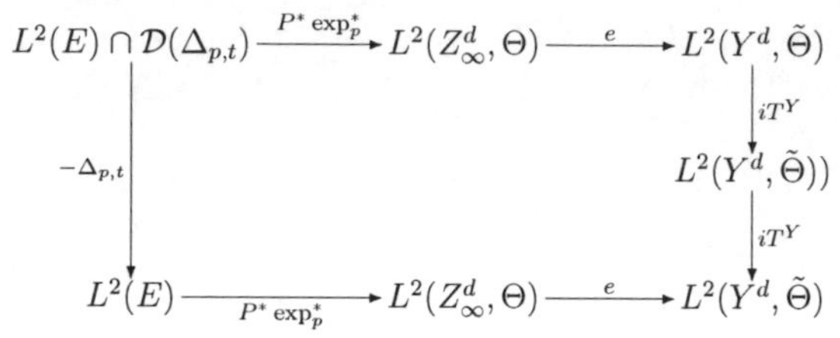

$$L^2(E) \cap \mathcal{D}(\Delta_{p,t}) \xrightarrow{\ P^* \exp_p^* \ } L^2(Z_\infty^d, \Theta) \xrightarrow{\quad e \quad} L^2(Y^d, \tilde{\Theta})$$

$$\downarrow{-\Delta_{p,t}} \qquad\qquad\qquad\qquad\qquad \downarrow{iT^Y}$$

$$\qquad\qquad\qquad\qquad\qquad\qquad L^2(Y^d, \tilde{\Theta}))$$

$$\qquad\qquad\qquad\qquad\qquad\qquad \downarrow{iT^Y}$$

$$L^2(E) \xrightarrow{\ P^* \exp_p^* \ } L^2(Z_\infty^d, \Theta) \xrightarrow{\quad e \quad} L^2(Y^d, \tilde{\Theta})$$

Figure 10: Commutative diagram for the decomposition of $-\Delta_{p,t}$ for Riemannian manifolds E diffeomorphic to $\mathbb{R}^d$.

2.5. Asymptotic sharpness of the uncertainty principles on compacta

In this section, we show that the uncertainty principle (2.94) for compact Riemannian manifolds M and the uncertainty principle (2.110) for compact star-shaped domains Ω are asymptotically sharp, i.e., that there exists a family of functions H_λ in the domain of the differential operator $\frac{\partial}{\partial t_*}$ on M and Ω such that for $\lambda \to 0$ equality is attained in (2.94) and (2.110), respectively. For this purpose, we construct a family H_λ of Gaussian-like functions on M and Ω.

First, we prove an auxiliary result. For $k \in \mathbb{N}_0$ and $\sigma > 0$, we have the following well-known moment formulas for the Gaussian function (cf. [64, p. 110]):

$$\int_0^\infty t^{2k} e^{-\frac{t^2}{\sigma^2}} \, dt = \frac{\sqrt{\pi}}{2} \frac{(2k)!}{4^k k!} \sigma^{2k+1}, \tag{2.124}$$

$$\int_0^\infty t^{2k+1} e^{-\frac{t^2}{\sigma^2}} \, dt = \frac{k!}{2} \sigma^{2k+2}. \tag{2.125}$$

On $[0,\infty)$, we introduce for $d \in \mathbb{N}$, $d \geq 1$ and $\sigma > 0$ the Gaussians $G_{d,\sigma}$ as

$$G_{d,\sigma}(t) := \begin{cases} \sqrt{\dfrac{2}{\sqrt{\pi}} \dfrac{4^k k!}{(2k)!}} \dfrac{1}{\sigma^{k+1/2}} e^{-\frac{t^2}{2\sigma^2}} & \text{if } d = 2k+1, \\[2ex] \sqrt{\dfrac{2}{k!}} \dfrac{1}{\sigma^{k+1}} e^{-\frac{t^2}{2\sigma^2}} & \text{if } d = 2k+2. \end{cases}$$

Then, the moment formulas (2.124) and (2.125) imply that $G_{d,\sigma}$ is a normalized function in the Hilbert space $L_d^2 := L^2([0,\infty), t^{d-1})$ with $\|G_{d,\sigma}\|_{L_d^2} = 1$. Moreover, the following properties hold:

Lemma 2.57.
Consider $G_{d,\sigma}$ as an element of the Hilbert space L_d^2. Then,

$$\|tG_{d,\sigma}\|_{L_d^2}^2 = \frac{d}{2}\sigma^2, \tag{2.126}$$

$$\|G'_{d,\sigma}\|_{L_d^2}^2 = \frac{d}{2}\frac{1}{\sigma^2}. \tag{2.127}$$

Proof. We prove equations (2.126) and (2.127) by direct calculation using the formulas (2.124) and (2.125). For d odd and $k = \frac{d-1}{2}$, we get

$$\|G'_{d,\sigma}\|^2_{L^2_d} = \int_0^\infty \frac{2}{\sqrt{\pi}} \frac{4^k k!}{(2k)!} \frac{1}{\sigma^{2k+1}} \left(\frac{d}{dt} e^{-\frac{t^2}{2\sigma^2}}\right)^2 t^{2k} dt$$

$$= \int_0^\infty \frac{2}{\sqrt{\pi}} \frac{4^k k!}{(2k)!} \frac{1}{\sigma^{2k+5}} e^{-\frac{t^2}{\sigma^2}} t^{2k+2} dt = \frac{1}{4} \frac{k!}{(2k)!} \frac{(2k+2)!}{(k+1)!} \frac{1}{\sigma^2} = \frac{d}{2} \frac{1}{\sigma^2}.$$

$$\|tG_{d,\sigma}\|^2_{L^2_d} = \int_0^\infty \frac{2}{\sqrt{\pi}} \frac{4^k k!}{(2k)!} \frac{1}{\sigma^{2k+1}} e^{-\frac{t^2}{\sigma^2}} t^{2k+2} dt = \frac{d}{2} \sigma^2.$$

On the other hand, for d even and $k = \frac{d-2}{2}$, we have

$$\|G'_{d,\sigma}\|^2_{L^2_d} = \int_0^\infty \frac{2}{k!} \frac{1}{\sigma^{2k+2}} \left(\frac{d}{dt} e^{-\frac{t^2}{2\sigma^2}}\right)^2 t^{2k} dt$$

$$= \int_0^\infty \frac{2}{k!} \frac{1}{\sigma^{2k+6}} e^{-\frac{t^2}{\sigma^2}} t^{2k+2} dt = \frac{(k+1)!}{k!} \frac{1}{\sigma^2} = \frac{d}{2} \frac{1}{\sigma^2}.$$

$$\|tG_{d,\sigma}\|^2_{L^2_d} = \int_0^\infty \frac{2}{k!} \frac{1}{\sigma^{2k+2}} e^{-\frac{t^2}{\sigma^2}} t^{2k+2} dt = \frac{d}{2} \sigma^2.$$

$\square$

Now, we consider first the uncertainty principle on a compact Riemannian manifold M proven in Section 2.2 and show that inequality (2.94) is asymptotically sharp. We start out by choosing $\delta > 0$ small enough such that the open ball $B(p, \delta)$ with center p and radius δ is a subset of $D_p \subset M$. Further, we define a C^∞-cut-off function $\varphi_\delta : [0, \infty) \to [0, 1]$ with the property that $\varphi_\delta(t) = 1$ for $0 \leq t \leq \frac{\delta}{2}$, $0 \leq \varphi_\delta(t) \leq 1$ for $\frac{\delta}{2} \leq t \leq \delta$, and $\varphi_\delta(t) = 0$ for $t \geq \delta$. Then, we set $c_\xi := \sqrt{\frac{R(\xi)}{\pi}}$ and define for $\lambda \in (0, \infty)$ the family of functions $\tilde{H}_\lambda$ in geodesic polar coordinates at $p \in M$ by

$$\tilde{H}^*_\lambda(t, \xi) := G_{d, c_\xi \lambda}(t) \varphi_\delta(t) \tag{2.128}$$

and its normalization as

$$H_\lambda := \frac{\tilde{H}_\lambda}{\|\tilde{H}_\lambda\|_M}. \tag{2.129}$$

Because of the cut-off function φ_δ, the functions H_λ are compactly supported in $B(p, \delta) \subset M$ and, in particular, elements of $L^2(M) \cap \mathcal{D}(\frac{\partial}{\partial t_*}; M)$. Now, we can show the following proposition:

Proposition 2.58.
Let $|\mathfrak{S}_p|$ denote the volume of the unit sphere $\mathfrak{S}_p$, then

$$\lim_{\lambda \to 0} \frac{1 - \varepsilon_p(H_\lambda)^2}{\lambda^2} = \frac{d}{2|\mathfrak{S}_p|} \int_{\mathfrak{S}_p} \frac{\pi}{R(\xi)} d\mu(\xi), \tag{2.130}$$

$$\lim_{\lambda \to 0} \lambda^2 \left\| \frac{\partial}{\partial t_*} H_\lambda \right\|_M^2 = \frac{d}{2|\mathfrak{S}_p|} \int_{\mathfrak{S}_p} \frac{\pi}{R(\xi)} d\mu(\xi), \tag{2.131}$$

$$\lim_{\lambda \to 0} \rho_p(H_\lambda) = \frac{d}{|\mathfrak{S}_p|} \int_{\mathfrak{S}_p} \frac{\pi}{R(\xi)} d\mu(\xi). \tag{2.132}$$

In particular, the uncertainty inequality (2.94) is asymptotically sharp.

Proof. Beside Lemma 2.57, we need two facts for the proof. The first one is a property of the weight function Θ. If $\delta > 0$ is chosen small enough, we have for $t \leq \delta$ the Taylor expansion (cf. (2.80) and [8, XII 8])

$$\Theta(t, \xi) = t^{d-1} - \frac{\mathrm{Ric}(\xi, \xi)}{6} t^{d+1} + \mathrm{O}(t^{d+2}), \tag{2.133}$$

$$\Theta'(t, \xi) = (d-1)t^{d-2} - \frac{(d+1)\,\mathrm{Ric}(\xi, \xi)}{6} t^d + \mathrm{O}(t^{d+1}), \tag{2.134}$$

where $\mathrm{Ric}(\cdot, \cdot)$ denotes the Ricci tensor on $T_p M \times T_p M$ (see Section A.6). The second fact concerns the Gaussian function $G_{d, c_\xi \lambda}$. Since the term $c_\xi = \sqrt{\frac{R(\xi)}{\pi}}$ is uniformly bounded above and below by positive constants, there exists for $\delta > 0$ and $\epsilon > 0$ a $\lambda_{\delta, \epsilon}$ such that for all $\lambda < \lambda_{\delta, \epsilon}$ and $\xi \in \mathfrak{S}_p$ we have

$$\int_{\delta/2}^\infty G_{d, c_\xi \lambda}(t)^2 t^{d-1} dt < \epsilon. \tag{2.135}$$

We consider now the L^2-norm of $\tilde{H}_\lambda$ on M. Using the Taylor expansion (2.133) of the weight function Θ and property (2.126) of Lemma 2.57, we get the estimate

$$\begin{aligned}
\lim_{\lambda \to 0} \|\tilde{H}_\lambda\|_M^2 &= \lim_{\lambda \to 0} \int_{\mathfrak{S}_p} \int_0^\delta G_{d, c_\xi \lambda}(t)^2 \varphi_\delta(t)^2 \Theta(t, \xi) dt d\mu(\xi) \\
&= \lim_{\lambda \to 0} \int_{\mathfrak{S}_p} \int_0^\delta G_{d, c_\xi \lambda}(t)^2 \varphi_\delta(t)^2 \Big(t^{d-1} + \mathrm{O}(t^{d+1}) \Big) dt d\mu(\xi) \\
&\leq \lim_{\lambda \to 0} \int_{\mathfrak{S}_p} \int_0^\infty G_{d, c_\xi \lambda}(t)^2 \Big(t^{d-1} + \mathrm{O}(t^{d+1}) \Big) dt d\mu(\xi) \\
&= \lim_{\lambda \to 0} |\mathfrak{S}_p| + \mathrm{O}(\lambda^2) = |\mathfrak{S}_p|,
\end{aligned}$$

where $|\mathfrak{S}_p|$ denotes the volume of the $(d-1)$-dimensional unit sphere $\mathfrak{S}_p$ in the tangent space $T_p M$. Using property (2.135), we get for an arbitrary $\epsilon > 0$ and $\lambda < \lambda_{\delta, \epsilon}$

$$\begin{aligned}
\|\tilde{H}_\lambda\|_M^2 &= \int_{\mathfrak{S}_p} \int_0^\delta G_{d, c_\xi \lambda}(t)^2 \varphi_\delta(t)^2 \Big(t^{d-1} + \mathrm{O}(t^{d+1}) \Big) dt d\mu(\xi) \\
&\geq \int_{\mathfrak{S}_p} \int_0^\infty G_{d, c_\xi \lambda}(t)^2 \Big(t^{d-1} + \mathrm{O}(t^{d+1}) \Big) dt d\mu(\xi) - \epsilon |\mathfrak{S}_p| \\
&= (1 - \epsilon) |\mathfrak{S}_p| + \mathrm{O}(\lambda^2).
\end{aligned}$$

Therefore,

$$\lim_{\lambda \to 0} \|\tilde{H}_\lambda\|_M^2 = |\mathfrak{S}_p|. \tag{2.136}$$

We consider now equation (2.130). Using the Taylor expansion (2.133) of the weight function Θ and property (2.126), we get the upper estimate

$$
\begin{aligned}
1 - \varepsilon_p(H_\lambda) &= \frac{1}{\|\tilde{H}_\lambda\|_M^2} \int_{\mathfrak{S}_p} \int_0^\delta \left(1 - \cos(\tfrac{\pi}{R(\xi)}t)\right) G_{d,c_\xi\lambda}(t)^2 \varphi_\delta(t)^2 \Theta(t,\xi) dt d\mu(\xi) \\
&\leq \frac{1}{\|\tilde{H}_\lambda\|_M^2} \int_{\mathfrak{S}_p} \int_0^\delta \tfrac{\pi^2}{R(\xi)^2} \tfrac{t^2}{2} G_{d,c_\xi\lambda}(t)^2 \varphi_\delta(t)^2 \Theta(t,\xi) dt d\mu(\xi) \\
&\leq \frac{1}{\|\tilde{H}_\lambda\|_M^2} \int_{\mathfrak{S}_p} \int_0^\infty \tfrac{\pi^2}{R(\xi)^2} \tfrac{t^2}{2} G_{d,c_\xi\lambda}(t)^2 \left(t^{d-1} + O(t^{d+1})\right) dt d\mu(\xi) \\
&= \frac{d}{4} \lambda^2 \frac{1}{\|\tilde{H}_\lambda\|_M^2} \int_{\mathfrak{S}_p} \tfrac{\pi}{R(\xi)} d\mu(\xi) + O(\lambda^4).
\end{aligned}
$$

Further, since $\|H_\lambda\|_M = 1$, we have $\varepsilon_p(H_\lambda) \leq 1$ and, hence, $(1 + \varepsilon_p(H_\lambda)) \leq 2$. In total, we get

$$
\lim_{\lambda \to 0} \frac{1 - \varepsilon_p(H_\lambda)^2}{\lambda^2} \leq 2 \lim_{\lambda \to 0} \frac{1 - \varepsilon_p(H_\lambda)}{\lambda^2} \leq \frac{d}{2|\mathfrak{S}_p|} \int_{\mathfrak{S}_p} \tfrac{\pi}{R(\xi)} d\mu(\xi). \tag{2.137}
$$

Next, we turn to equation (2.131). For the following estimate, we use the Taylor expansion (2.133) and equation (2.127) of Lemma 2.57.

$$
\begin{aligned}
\left\|\frac{\partial}{\partial t_*} H_\lambda\right\|_M^2 &= \frac{1}{\|\tilde{H}_\lambda\|_M^2} \int_{\mathfrak{S}_p} \int_0^\delta \left|\frac{\partial}{\partial t}\left(G_{d,c_\xi\lambda}(t)\varphi_\delta(t)\right)\right|^2 \Theta(t,\xi) dt d\mu(\xi) \\
&\leq \frac{1}{\|\tilde{H}_\lambda\|_M^2} \int_{\mathfrak{S}_p} \int_0^\infty \left[|G'_{d,c_\xi\lambda}(t)|^2 + 2|G'_{d,c_\xi\lambda}(t)|\|\varphi'_\delta\|_\infty + \right. \\
&\qquad\qquad \left. + |G_{d,c_\xi\lambda}(t)|\|\varphi'_\delta\|_\infty^2\right]\left(t^{d-1} + O(t^{d+1})\right) dt d\mu(\xi) \\
&= \frac{d}{2} \frac{1}{\lambda^2} \frac{1}{\|\tilde{H}_\lambda\|_M^2} \int_{\mathfrak{S}_p} \tfrac{\pi}{R(\xi)} d\mu(\xi) + O\left(\tfrac{1}{\lambda}\right).
\end{aligned}
$$

Thus, we get

$$
\lim_{\lambda \to 0} \lambda^2 \left\|\frac{\partial}{\partial t_*} H_\lambda\right\|_M^2 \leq \frac{d}{2|\mathfrak{S}_p|} \int_{\mathfrak{S}_p} \tfrac{\pi}{R(\xi)} d\mu(\xi). \tag{2.138}
$$

Finally, we take a look at equation (2.132). Due to (2.133) and (2.134), the function $\frac{\Theta'(t,\xi)}{\Theta(t,\xi)} \sin(\tfrac{\pi}{R(\xi)}t)$ has for small t the Taylor expansion

$$
\frac{\Theta'(t,\xi)}{O(t,\xi)} \sin(\tfrac{\pi}{R(\xi)}t) = (d-1)\frac{\pi}{R(\xi)} + O(t^2). \tag{2.139}
$$

Using (2.139) and property (2.135), we derive for an arbitrary $\epsilon > 0$ and for $\lambda < \lambda_{\delta,\epsilon}$

$$
\begin{aligned}
\rho_p(H_\lambda) &= \int_{\mathbb{S}_p} \int_0^\delta \left(\frac{\pi}{R(\xi)} \cos(\frac{\pi}{R(\xi)}t) + \frac{\Theta(t,\xi)'}{\Theta(t,\xi)} \sin(\frac{\pi}{R(\xi)}t) \right) |H_\lambda^*(t,\xi)|^2 \Theta(t,\xi) dt d\mu(\xi) \\
&= \frac{1}{\|\tilde{H}_\lambda\|_M^2} \int_{\mathbb{S}_p} \int_0^\delta \left(\frac{\pi}{R(\xi)} \cos(\frac{\pi}{R(\xi)}t) + \frac{\Theta(t,\xi)'}{\Theta(t,\xi)} \sin(\frac{\pi}{R(\xi)}t) \right) \\
&\qquad\qquad \times \left(G_{d,c_\xi\lambda}(t)\varphi_\delta(t) \right)^2 \Theta(t,\xi) dt d\mu(\xi) \\
&= \frac{d}{\|\tilde{H}_\lambda\|_M^2} \int_{\mathbb{S}_p} \frac{\pi}{R(\xi)} \int_0^\delta \left(G_{d,c_\xi\lambda}(t)\varphi_\delta(t) \right)^2 \left(t^{d-1} + O(t^{d+1}) \right) dt d\mu(\xi) \\
&\geq \frac{d}{\|\tilde{H}_\lambda\|_M^2} \int_{\mathbb{S}_p} \frac{\pi}{R(\xi)} d\mu(\xi)(1 - \epsilon) + O(\lambda^2).
\end{aligned}
$$

Thus, we conclude

$$
\lim_{\lambda \to 0} \rho_p(H_\lambda) \geq \frac{d}{|\mathbb{S}_p|} \int_{\mathbb{S}_p} \frac{\pi}{R(\xi)} d\mu(\xi). \tag{2.140}
$$

Now, inserting the inequalities (2.137), (2.138) and (2.140) in the uncertainty inequality (2.90), we get the same value on both sides, namely $\frac{d^2}{4|\mathbb{S}_p|^2} (\int_{\mathbb{S}_p} \frac{\pi}{R(\xi)} d\mu(\xi))^2$. Thus, inequalities (2.137), (2.138) and (2.140) are in fact equalities and the proposition is proven. $\square$

Similarly, if Ω is a compact star-shaped subset of M, it is possible to prove that the uncertainty principle (2.110) is asymptotically sharp. As above, we choose $\delta > 0$ small enough such that $B(p,\delta) \subset \Omega$ and use φ_δ as a cut-off-function. Similar as in (2.128), we define for $\lambda \in (0,\infty)$ a Gaussian-type family of functions $\tilde{H}_\lambda^\Omega$ by

$$
\tilde{H}_\lambda^{\Omega*}(t,\xi) := G_{d,\lambda}(t)\varphi_\delta(t)
$$

and the normalized functions by

$$
H_\lambda^\Omega := \frac{\tilde{H}_\lambda^\Omega}{\|\tilde{H}_\lambda^\Omega\|_\Omega}.
$$

Obviously, the function H_λ^Ω is supported in $B(p,\delta)$ and is an element of $L^2(M) \cap \mathcal{D}_0(\frac{\partial}{\partial t_*}; \Omega)$. Moreover, we get the following asymptotic result:

Proposition 2.59.

$$
\lim_{\lambda \to 0} \frac{\|t_* H_\lambda^\Omega\|_\Omega^2}{\lambda^2} = \frac{d}{2}, \tag{2.141}
$$

$$
\lim_{\lambda \to 0} \lambda^2 \left\| \frac{\partial}{\partial t_*} H_\lambda^\Omega \right\|_\Omega^2 = \frac{d}{2}, \tag{2.142}
$$

$$
\lim_{\lambda \to 0} 1 + \int_{\mathbb{S}_p} \int_0^{Q(\xi)} t\Theta'(t,\xi)) |H_\lambda^{\Omega*}(t,\xi)|^2 dt d\mu(\xi) = d. \tag{2.143}
$$

In particular, the uncertainty principle (2.110) is asymptotically sharp.

Proof. We consider first the L^2-norm of $\tilde{H}_\lambda^\Omega$ on Ω. Similar as in (2.136), we get the asymptotic formula

$$\lim_{\lambda \to 0} \|\tilde{H}_\lambda^\Omega\|_\Omega^2 = |\mathfrak{S}_p|. \tag{2.144}$$

Next, we turn to equation (2.141). Using the Taylor expansion (2.133) of the weight function Θ and property (2.126), we get the upper bound

$$\|t_* H_\lambda^\Omega\|_\Omega^2 = \frac{1}{\|\tilde{H}_\lambda^\Omega\|_\Omega^2} \int_{\mathfrak{S}_p} \int_0^\delta t^2 G_{d,\lambda}(t)^2 \varphi_\delta(t)^2 \Theta(t,\xi) dt d\mu(\xi)$$

$$\leq \frac{1}{\|\tilde{H}_\lambda^\Omega\|_\Omega^2} \int_{\mathfrak{S}_p} \int_0^\infty G_{d,\lambda}(t)^2 \left(t^{d+1} + O(t^{d+3})\right) dt d\mu(\xi)$$

$$= \frac{d}{2} \lambda^2 \frac{|\mathfrak{S}_p|}{\|\tilde{H}_\lambda^\Omega\|_\Omega^2} + O(\lambda^4).$$

Thus, in the limit $\lambda \to 0$, we get

$$\lim_{\lambda \to 0} \frac{\|t_* H_\lambda^\Omega\|_\Omega^2}{\lambda^2} \leq \frac{d}{2}. \tag{2.145}$$

Next, we consider equation (2.142). Proceeding in the same way as in the proof of inequality (2.138), we get the estimate

$$\lim_{\lambda \to 0} \lambda^2 \left\|\frac{\partial}{\partial t_*} H_\lambda^\Omega\right\|_\Omega^2 \leq \frac{d}{2}. \tag{2.146}$$

Finally, we turn to equation (2.143). Due to (2.134) and property (2.135), we derive for an arbitrary $\epsilon > 0$ and for $\lambda < \lambda_{\delta,\epsilon}$

$$\int_{\mathfrak{S}_p} \int_0^{Q(\xi)} t\Theta'(t,\xi))|H_\lambda^\Omega(t,\xi)|^2 dt d\mu(\xi) =$$

$$= \frac{1}{\|\tilde{H}_\lambda\|_\Omega^2} \int_{\mathfrak{S}_p} \int_0^\delta t\Theta'(t,\xi)) \left(G_{d,\lambda}(t)\varphi_\delta(t)\right)^2 dt d\mu(\xi)$$

$$\geq \frac{d-1}{\|\tilde{H}_\lambda\|_\Omega^2} \int_{\mathfrak{S}_p} \left(\int_0^\infty \left(G_{d,\lambda}(t)\varphi_\delta(t)\right)^2 \left(t^{d-1} + O(t^{d+1})\right) dt - \epsilon \right) d\mu(\xi)$$

$$\geq \frac{(d-1)|\mathfrak{S}_p|}{\|\tilde{H}_\lambda^\Omega\|_\Omega^2} (1-\epsilon) + O(\lambda^2).$$

Thus, we can conclude

$$\lim_{\lambda \to 0} \left(1 + \int_{\mathfrak{S}_p} \int_0^{Q(\xi)} t\Theta'(t,\xi))|H_\lambda^{\Omega*}(t,\xi)|^2 dt d\mu(\xi)\right) \geq d. \tag{2.147}$$

Using the inequalities (2.145), (2.146) and (2.147) in the uncertainty inequality (2.110), we get on both sides the value $\frac{d^2}{4}$. Thus, we have shown that the inequalities (2.145), (2.146) and (2.147) are in fact equalities. □

2.6. Examples

2.6.1. The spheres

As a first example of a compact Riemannian manifold, we start with the d-dimensional sphere

$$\mathbb{S}_r^d = \{q \in \mathbb{R}^{d+1} : q_1^2 + \ldots + q_{d+1}^2 = r^2\}$$

with radius $r > 0$. The sphere $\mathbb{S}_r^d$ is a submanifold of $\mathbb{R}^{d+1}$ and the canonical Riemannian structure on $\mathbb{S}_r^d$ is defined by the restriction of the standard Euclidean metric in $\mathbb{R}^{d+1}$ to the submanifold $\mathbb{S}_r^d$. If $p \in \mathbb{S}_r^d$, we identify the tangent space $T_p\mathbb{S}_r^d$ with the orthogonal complement $p^\perp$ of the linear vector space $\mathbb{R}p$ in $\mathbb{R}^{d+1}$. An arbitrary point $q \in \mathbb{S}_r^d$ can then be represented as

$$q = q(t, \xi) = r\cos(\tfrac{t}{r})p + r\sin(\tfrac{t}{r})\xi,$$

where $t \in [0, r\pi]$ and $\xi \in \mathfrak{S}_p$ is a unit vector in the hyperplane $p^\perp$. For fixed ξ, the functions $\gamma_\xi(t) = q(t, \xi)$ describe the geodesics on $\mathbb{S}_r^d$ starting at $\gamma_\xi(0) = p$ (see [9, Section II.3]), and the coordinates (t, ξ) correspond to the geodesic polar coordinates at p. The cut locus C_p consists of the single point $\{-p\}$ lying at the antipodal end of the sphere. Further, the distance value $R(\xi)$ is, independently from the directional variable ξ, equal to $r\pi$. The weight function Θ can be determined as (cf. [3, p. 57])

$$\Theta(t, \xi) = r^{d-1}\sin^{d-1}(\tfrac{t}{r}), \tag{2.148}$$

and the Laplace-Beltrami operator on $\mathbb{S}_r^d$ as (cf. [8, II.5, equation (29)])

$$(\Delta_{\mathbb{S}_r^d}f)^*(t, \xi) = \frac{\partial^2}{\partial t}f^*(t, \xi) + \frac{d-1}{r}\cot(\tfrac{t}{r})\frac{\partial}{\partial t}f^*(t, \xi) + \frac{\Delta_{\mathfrak{S}_p}(f^*(t, \xi)|_{\mathfrak{S}_p})}{r^2\sin^2(\tfrac{t}{r})}, \tag{2.149}$$

for $t \in (0, \pi)$ and $\xi \in \mathfrak{S}_p$. The radial part of the Laplacian is given as

$$(\Delta_{p,t}f)^*(t, \xi) = \frac{\partial^2}{\partial t}f^*(t, \xi) + \frac{d-1}{r}\cot(\tfrac{t}{r})\frac{\partial}{\partial t}f^*(t, \xi). \tag{2.150}$$

Now, an uncertainty principle on $\mathbb{S}_r^d$ can be formulated as follows.

Corollary 2.60.
Let $p \in \mathbb{S}_r^d$ and $f \in L^2(\mathbb{S}_r^d) \cap \mathcal{D}(\frac{\partial}{\partial t_*}; \mathbb{S}_r^d)$ be normalized such that $\|f\|_{\mathbb{S}_r^d} = 1$ and $\varepsilon_p(f) \neq 0$. Then, the following uncertainty principle holds:

$$r^2\frac{1 - \varepsilon_p(f)^2}{\varepsilon_p(f)^2} \cdot \left\|\frac{\partial}{\partial t_*}f\right\|_{\mathbb{S}_r^d}^2 > \frac{d^2}{4}. \tag{2.151}$$

The constant $\frac{d^2}{4}$ is optimal.

Proof. If we apply Theorem 2.41 to the sphere $\mathbb{S}_r^d$ and use the respective weight function (2.148), the only thing that remains to validate is the right hand side of inequality (2.90). This is done by the following simple calculation.

$$\rho_p(f) = \int_{\mathfrak{S}_p} \int_0^{r\pi} \left(\frac{1}{r}\cos(\frac{t}{r}) + \frac{\Theta'(t,\xi)}{\Theta(t,\xi)}\sin(\frac{t}{r}) \right) |f^*(t,\xi)|^2 \Theta(t,\xi) dt d\mu(\xi)$$

$$= \frac{1}{r}\varepsilon_p(f) + \frac{d-1}{r}\int_{\mathfrak{S}_p}\int_0^{r\pi} \cot(\frac{t}{r})\sin(\frac{t}{r})|f^*(t,\xi)|^2 \sin^{d-1}(\frac{t}{r}) dt d\mu(\xi) = \frac{d}{r}\varepsilon_p(f).$$

The optimality of the constant $\frac{d^2}{4}$ is a direct consequence of Proposition 2.58. $\qquad\square$

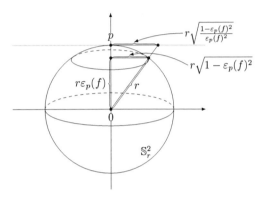

Figure 11: Geometric interpretation of the position variance $\mathrm{var}_{\mathbb{S},p}^{\mathbb{S}_r^2}(f)$ on the sphere $\mathbb{S}_r^2$ with radius $r > 0$.

If we consider only the radial functions on the unit sphere $\mathbb{S}^d = \mathbb{S}_1^d$, inequality (2.151) corresponds exactly with the uncertainty principle (1.73) proven in [73] for functions having an expansion in terms of the Gegenbauer polynomials $C_n^{(\frac{d-1}{2})}$. This is not surprising since the polynomials $C_n^{(\frac{d-1}{2})}$ constitute a basis for the radial, square integrable functions on $\mathbb{S}^d$ and also the radial Laplacian (2.150) corresponds to the second-order differential operator $L_{\frac{d-2}{2},\frac{d-2}{2}}$ of the corresponding Gegenbauer polynomials defined in (1.67). The sharpness of inequality (2.151) is therefore also a consequence of the sharpness of the uncertainty inequality (1.73) and vice versa.

2.6.2. The projective spaces

Our next main examples are the projective spaces. We start with the d-dimensional real projective space $\mathbb{RP}_r^d$. We consider the standard sphere $\mathbb{S}_{2r}^d$ with radius $2r$ and define the antipodal map $A : \mathbb{S}_{2r}^d \to \mathbb{S}_{2r}^d$ by $Ap = -p$. The real projective space $\mathbb{RP}_r^d$ with diameter

$r\pi$ is then defined as the quotient of $\mathbb{S}_{2r}^d$ under the group $G = \{\mathrm{id}, A\}$. Since G is a proper and free isometry group on $\mathbb{S}_{2r}^d$, we get in a canonical way a Riemannian metric on the quotient $\mathbb{RP}_r^d$ from the standard Riemannian metric on the covering $\mathbb{S}_{2r}^d$. Moreover, the identification of $\mathbb{RP}_r^d$ with $\mathbb{S}_{2r}^d/G$ allows the introduction of geodesic polar coordinates at a point p as in the case of the sphere. In this way, the Riemannian measure on $\mathbb{RP}_r^d$ can be deduced from (2.148) as

$$d\mu^*_{\mathbb{RP}_r^d}(t,\xi) = (2r)^{d-1}\sin^{d-1}(\tfrac{t}{2r})dt d\mu(\xi), \quad (t,\xi) \in [0,r\pi] \times \mathfrak{S}_p. \tag{2.152}$$

The cut locus C_p on $\mathbb{RP}_r^d$ corresponds to the set of points lying on the equator of $\mathbb{S}_{2r}^d$ with respect to the pole p. Since the antipodal points are also identified with each other on the equator, the cut locus C_p is isometric to the real projective space $\mathbb{RP}_r^{d-1}$. Due to our special construction, the distance $R(\xi)$ from p to the cut locus is, independently of the direction ξ, equal to $r\pi$. Further, the Laplace-Beltrami operator on $\mathbb{RP}_r^d$ can be deduced from (2.149) as

$$(\Delta_{\mathbb{RP}_r^d}f)^*(t,\xi) = \frac{\partial^2 f^*}{\partial t^2}(t,\xi) + \frac{(d-1)}{2r}\frac{1+\cos(\tfrac{t}{r})}{\sin(\tfrac{t}{r})}\frac{\partial f^*}{\partial t}(t,\xi) + \frac{\Delta_{\mathfrak{S}_p}(f^*(t,\xi)|_{\mathfrak{S}_p})}{4r^2\sin^2(\tfrac{t}{2r})}. \tag{2.153}$$

Next, we take a look at the complex projective space $\mathbb{CP}_r^d$, $d = 2, 4, 6, \ldots$. We identify the complex space $\mathbb{C}^{\frac{d}{2}+1}$ with the real Euclidean space $\mathbb{R}^{d+2}$ and consider the sphere $\mathbb{S}_{2r}^{d+1}$ as a Riemannian submanifold of $\mathbb{C}^{\frac{d}{2}+1}$. Then, the complex projective space $\mathbb{CP}_r^d$ is defined as the quotient of $\mathbb{S}_{2r}^{d+1} \subset \mathbb{C}^{\frac{d}{2}+1}$ under the group of complex scalars of absolute value 1 acting on $\mathbb{S}_{2r}^{d+1} \subset \mathbb{C}^{\frac{d}{2}+1}$. The projection $\mathbb{S}_{2r}^{d+1} \to \mathbb{CP}_r^d$ is a fibration (in particular a submersion, see [3, Chapter 1, E5] and Section A.1 of the Appendix) known as the Hopf fibration, and $\mathbb{CP}_r^d$ can therefore be endowed with a unique Riemannian structure. If we introduce geodesic polar coordinates $(t,\xi) \in [0,r\pi] \times \mathfrak{S}_p$ at a point $p \in \mathbb{CP}_r^d$, then the weight function $\Theta(t,\xi)$ assumes, independently of the directional variable ξ, the value (cf. [3, Chapter 2, F42], [35, p. 171])

$$\Theta(t,\xi) = (2r)^{d-1}\sin^{d-1}(\tfrac{t}{2r})\cos(\tfrac{t}{2r}), \tag{2.154}$$

and the geodesic distance $R(\xi)$ from p to the cut locus $C_p \simeq \mathbb{CP}_r^{d-2}$ is equal to $r\pi$. So, the radial Laplacian $\Delta_{p,t}$ reads as

$$(\Delta_{p,t}f)^*(t,\xi) = \frac{\partial^2 f^*}{\partial t^2}(t,\xi) + \frac{d-2+d\cos(\tfrac{t}{r})}{2r\sin(\tfrac{t}{r})}\frac{\partial f^*}{\partial t}(t,\xi). \tag{2.155}$$

Similarly, one obtains the quaternionic projective space $\mathbb{HP}_r^d$, $d = 4, 8, 12, \ldots$, by starting with the unit sphere $\mathbb{S}_{2r}^{d+3}$ as a Riemannian submanifold of the quaternionic space $\mathbb{H}^{\frac{d}{4}+1}$. Then $\mathbb{HP}_r^d$ is defined as the quotient of $\mathbb{S}_{2r}^{d+3}$ under the group of unit quaternions acting on $\mathbb{S}_{2r}^{d+3} \subset \mathbb{H}^{\frac{d}{4}+1}$. Again, the projection $\mathbb{S}_{2r}^{d+3} \to \mathbb{HP}_r^d$ is a fibration and $\mathbb{HP}_r^d$ can be endowed with a unique Riemannian structure. In geodesic polar coordinates at a point

p, the weight function Θ for the quaternionic complex space can be computed as (see [3, Chapter 2, F46], [35, p. 171])

$$\Theta(t,\xi) = (2r)^{d-1}\sin^{d-1}(\tfrac{t}{2r})\cos^3(\tfrac{t}{2r}), \qquad (2.156)$$

where $(t,\xi) \in [0,r\pi] \times \mathfrak{S}_p$. The cut locus at $t = R(\xi) = r\pi$ is a Riemannian submanifold of $\mathbb{HP}_r^d$ isometric to $\mathbb{HP}_r^{d-4}$. Moreover, the radial Laplacian $\Delta_{p,t}$ on $\mathbb{HP}_r^d$ reads as

$$(\Delta_{p,t}f)^*(t,\xi) = \frac{\partial^2 f^*}{\partial t^2}(t,\xi) + \frac{d-4+(d+2)\cos(\tfrac{t}{r})}{2r\sin(\tfrac{t}{r})}\frac{\partial f^*}{\partial t}(t,\xi). \qquad (2.157)$$

Beside the projective spaces $\mathbb{RP}_r^d$, $\mathbb{CP}_r^d$ and $\mathbb{HP}_r^d$, there exists a further projective space constructed upon the octonions, the so called Cayley plane $\mathbb{C}a_r$. As a 16-dimensional homogeneous space, it can be defined as the quotient of the exceptional Lie group $F_{4,(-52)}$ and the spin group $SO(9)$ (see [4, Chapter 3 G]). The weight function $\Theta(t,\xi)$ for the Cayley plane can be written as (see [3, p. 113], [35, p. 171])

$$\Theta(t,\xi) = (2r)^{15}\sin^{15}(\tfrac{t}{2r})\cos^7(\tfrac{t}{2r}), \qquad (2.158)$$

where $(t,\xi) \in [0,r\pi] \times \mathfrak{S}_p$. The cut locus C_p at $t = R(\xi) = r\pi$ is isometric to the sphere $\mathbb{S}_r^8$. The radial Laplacian $\Delta_{p,t}$ on $\mathbb{C}a_r$ reads as

$$\Delta_{p,t}f(t,\xi) = \frac{\partial^2}{\partial t^2}f^*(t,\xi) + \frac{4+11\cos(\tfrac{t}{r})}{r\sin(\tfrac{t}{r})}\frac{\partial}{\partial t}f^*(t,\xi). \qquad (2.159)$$

To unify the notation for the projective spaces, we set $\beta_M = -\tfrac{1}{2},0,1,3$ if $M = \mathbb{RP}_r^d, \mathbb{CP}_r^d, \mathbb{HP}_r^d, \mathbb{C}a_r$, respectively. Then, the weight function $\Theta_M(t,\xi)$ for the projective spaces can be written as

$$\Theta_M(t,\xi) = (2r)^{d-1}\sin^{d-1}(\tfrac{t}{2r})\cos^{2\beta_M+1}(\tfrac{t}{2r}), \qquad (2.160)$$

and the radial Laplacian reads as

$$(\Delta_{p,t}^M f)^*(t,\xi) = \frac{\partial^2}{\partial t^2}f^*(t,\xi) + \frac{d-2-2\beta_M+(d+2\beta_M)\cos(\tfrac{t}{r})}{2r\sin(\tfrac{t}{r})}\frac{\partial}{\partial t}f^*(t,\xi). \qquad (2.161)$$

So, an uncertainty principle for the projective spaces can be formulated as follows.

Corollary 2.61.
Let M be one of the projective spaces $M = \mathbb{RP}_r^d, \mathbb{CP}_r^d, \mathbb{HP}_r^d$ and $\mathbb{C}a_r$. Let $p \in M$ and $f \in L^2(M) \cap \mathcal{D}(\tfrac{\partial}{\partial t_*}; M)$ such that $\|f\|_M = 1$ and

$$\rho_p(f) = \frac{d-2-2\beta_M}{2r} + \frac{d+2+2\beta_M}{2r}\varepsilon_p(f) \neq 0.$$

Then, the following uncertainty principle holds:

$$r^2\frac{1-\varepsilon_p(f)^2}{\left(\frac{d-2-2\beta_M}{2d}+\frac{d+2+2\beta_M}{2d}\varepsilon_p(f)\right)^2}\cdot\left\|\frac{\partial}{\partial t_*}f\right\|_M^2 > \frac{d^2}{4}. \qquad (2.162)$$

The constant $\frac{d^2}{4}$ on the right hand side is optimal.

Proof. Due to equation (2.160), we have $\Theta_M(t,\xi) = (2r)^{d-1} \sin(\frac{t}{2r})^{d-1} \cos(\frac{t}{2r})^{2\beta_M+1}$. Now, if we apply Theorem 2.41, we get on the right hand side of inequality (2.90):

$$\rho_p(f) = \int_{\mathbb{S}_p} \int_0^{r\pi} \left(\frac{1}{r}\cos(\frac{t}{r}) + \frac{\Theta'_M(t,\xi)}{\Theta_M(t,\xi)}\sin(\frac{t}{r})\right)|f^*(t,\xi)|^2\Theta_M(t,\xi)dtd\mu(\xi)$$

$$= \frac{1}{r}\varepsilon_p(f) + \int_{\mathbb{S}_p}\int_0^{r\pi} \frac{d-2-2\beta_M + (d+2\beta_M)\cos(\frac{t}{r})}{2r}|f^*(t,\xi)|^2\Theta_M(t,\xi)dtd\mu(\xi)$$

$$= \frac{d-2-2\beta_M}{2r} + \frac{d+2+2\beta_M}{2r}\varepsilon_p(f).$$

The optimality of (2.162) follows from Proposition 2.58. $\qquad\square$

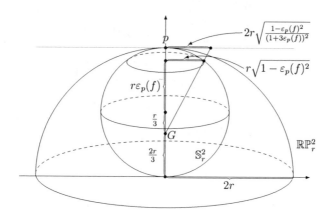

Figure 12: Geometric interpretation of the position variance $\mathrm{var}_{\mathbb{S},p}^{\mathbb{RP}_r^2}(f)$ for the real projective space $\mathbb{RP}_r^2$ with diameter $r\pi > 0$. The point G denotes the center of gravity of $\mathbb{RP}_r^2 \subset \mathbb{R}^3$.

It is well-known (see [24], [35]) that the radial square integrable functions on the projective space at a point p have an expansion in terms of the Jacobi polynomials $P_l^{(\frac{d-2}{2},\beta_M)}(\cos(\frac{t}{r}))$. Moreover, if $r = 1$, the radial Laplacian $\Delta_{p,t}^M$ in (3.49) corresponds exactly to the second-order differential operator $L_{\frac{d-2}{2},\beta_M}$ of the corresponding Jacobi polynomials $P_l^{(\frac{d-2}{2},\beta_M)}$ defined in (1.67). Thus, if $r = 1$, inequality (2.162) restricted to radial functions on the projective space is the same as the uncertainty principle (1.34) proven by Li and Liu in [54] for the Jacobi polynomials $P_l^{(\frac{d-2}{2},\beta_M)}$.

2.6.3. The flat tori

The flat tori $\mathbb{T}_r^d$ are defined as the quotients of the Euclidean space $\mathbb{R}^d$ under the action of the free abelian groups $(2r\mathbb{Z})^d$, $r > 0$. In this way, the flat tori $\mathbb{T}_r^d$ are compact Riemannian manifolds with curvature $\mathrm{K} = 0$. If $p \in \mathbb{T}_r^d$, the cut locus $\mathfrak{C}_p$ in the tangent space $T_p\mathbb{T}_r^d$ corresponds to the surface of the d-dimensional cuboid $[-r,r]^d$ with center 0 and edge length $2r$ (see also Example 2.33). Since $\mathbb{T}_r^d$ is a flat Riemannian manifold, the weight function Θ reads as

$$\Theta(t,\xi) = t^{d-1} \tag{2.163}$$

and the Laplace-Beltrami operator $\Delta_{\mathbb{T}_r^d}$ corresponds locally to the Laplacian in $\mathbb{R}^d$. If $q \in [-r,r]^d$, we set the Euclidean norm as $|q|^2 = t^2 = q_1^2 + \cdots + q_d^2$ and the maximum norm as $|q|_\infty = \max_{i=1,\ldots,d} |q_i|$. The distance function R can then be expressed in terms of $q \neq 0$ as $R(\frac{q}{|q|}) = r\frac{|q|}{|q|_\infty}$. The inverse coordinate transform $PL_R^{-1}P^{-1}$, which maps the cuboid $[-r,r]^d \setminus \{0\}$ onto the ball $B_\pi^d \setminus \{0\}$, can be written as $PL_R^{-1}P^{-1} : q \to \frac{\pi}{r}\frac{|q|_\infty}{|q|}q$. An uncertainty principle on $\mathbb{T}_r^d$ can now be formulated as follows.

Corollary 2.62.
Let $p \in \mathbb{T}_r^d$ and $f \in L^2(\mathbb{T}_r^d) \cap \mathcal{D}(\frac{\partial}{\partial t_*}; \mathbb{T}_r^d)$ be normalized such that $\|f\|_{\mathbb{T}_r^d} = 1$ and $\rho_p(f) \neq 0$. Then, the following uncertainty principle holds:

$$\frac{1 - \left(\int_{[-r,r]^d} \cos(\frac{\pi}{r}|q|_\infty)|f(p+q)|^2 dq\right)^2}{\left(\int_{[-r,r]^d} \left(\frac{\pi}{rd}\frac{|q|_\infty}{|q|}\cos(\frac{\pi}{r}|q|_\infty) + \frac{d-1}{d|q|}\sin(\frac{\pi}{r}|q|_\infty)\right)|f(p+q)|^2 dq\right)^2} \cdot \left\|\frac{\partial}{\partial t_*}f\right\|_{\mathbb{T}_r^d}^2 > \frac{d^2}{4}. \tag{2.164}$$

The constant $\frac{d^2}{4}$ is optimal.

Proof. We adopt Theorem 2.41, to derive an uncertainty principle for the flat torus $\mathbb{T}_r^d$. Clearly, the conditions of Theorem 2.41 are fulfilled. The value $\varepsilon_p(f)$ can be written as

$$\varepsilon_p(f) = \int_{\mathfrak{S}_p} \int_0^{R(\xi)} \cos(\tfrac{\pi}{R(\xi)}t)|f^*(t,\xi)|^2 t^{d-1} dt d\mu(\xi) = \int_{[-r,r]^d} \cos(\tfrac{\pi}{r}|q|_\infty)|f(p+q)|^2 dq.$$

Moreover, since $R(\frac{q}{|q|}) = r\frac{|q|}{|q|_\infty}$ and $|q| = t$, we get on the right hand side of (2.90)

$$\rho_p(f) = \int_{\mathfrak{S}_p} \int_0^{R(\xi)} \left(\tfrac{\pi}{R(\xi)}\cos(\tfrac{\pi}{R(\xi)}t) + \tfrac{\Theta'(t,\xi)}{\Theta(t,\xi)}\sin(\tfrac{\pi}{R(\xi)}t)\right)|f^*(t,\xi)|^2\Theta(t,\xi) dt d\mu(\xi)$$

$$= \int_{[-r,r]^d} \left(\tfrac{\pi}{r}\tfrac{|q|_\infty}{|q|}\cos(\tfrac{\pi}{r}|q|_\infty) + \tfrac{d-1}{|q|}\sin(\tfrac{\pi}{r}|q|_\infty)\right)|f(p+q)|^2 dq.$$

Thus, by Theorem 2.41, we get inequality (2.164) and the optimality of the constant $\frac{d^2}{4}$ follows from Proposition 2.58. $\square$

2.6.4. The hyperbolic spaces

As an example of a Riemannian manifold that is diffeomorphic to $\mathbb{R}^d$, we consider the hyperbolic space. If $\mathbb{R}^{d+1}$ is equipped with the symmetric bilinear form $(x, y) = -x_0 y_0 + \sum_{i=1}^{d} x_i y_i$, the hyperbolic space $\mathbb{H}_r^d$, $r > 0$, is defined as the submanifold of $\mathbb{R}^{d+1}$ satisfying

$$\mathbb{H}_r^d = \left\{ x \in \mathbb{R}^{d+1} : \ (x, x) = -r^2, \ x_0 > 0 \right\}.$$

The bilinear form $(\cdot, \cdot)$ restricted to $\mathbb{H}_r^d$ induces a positive definite metric on $\mathbb{H}_r^d$ and makes it to a Riemannian manifold. Moreover, it is well-known (see [9, Chapter II.3]) that the hyperbolic space $\mathbb{H}_r^d$ has constant negative sectional curvature $K = -\frac{1}{r^2}$ and the weight function Θ at a point $p \in \mathbb{H}_r^d$ is given by

$$\Theta(t, \xi) = r^{d-1} \sinh^{d-1}(\tfrac{t}{r}), \quad (t, \xi) \in [0, \infty) \times \mathfrak{S}_p, \tag{2.165}$$

$$\Theta'(t, \xi) = (d-1)r^{d-2} \cosh(\tfrac{t}{r}) \sinh^{d-2}(\tfrac{t}{r}). \tag{2.166}$$

Now, using Theorem 2.54, we get the following uncertainty principle for the hyperbolic space:

Corollary 2.63.
Let $p \in \mathbb{H}_r^d$ and $f \in L^2(\mathbb{H}_r^d)$, $\|f\|_{\mathbb{H}_r^d} = 1$, such that $f^* \in \mathcal{D}(\frac{\partial}{\partial t}, t, t\frac{\partial}{\partial t}; Z_\infty^d)$ and the consistency condition $f^*(0, \xi_1) = f^*(0, \xi_2)$ is satisfied for μ-a.e. $\xi_1, \xi_2 \in \mathfrak{S}_p$. Then, the following uncertainty principle holds:

$$\|t_* f\|_{\mathbb{H}_r^d}^2 \cdot \left\| \frac{\partial}{\partial t_*} f \right\|_{\mathbb{H}_r^d}^2 \geq \frac{1}{4} \left(1 + (d-1) \left\| \sqrt{\tfrac{t}{r} \coth(\tfrac{t}{r})}_* f \right\|_{\mathbb{H}_r^d}^2 \right)^2. \tag{2.167}$$

Equality in (2.167) is obtained if and only if $f^*(t, \xi) = Ce^{-\lambda t^2}$ for a nonnegative constant $\lambda > 0$ and a complex scalar C.

2.6.5. One-dimensional closed curves

If the manifold M is a closed one-dimensional curve, we can simplify inequality (2.90) considerably. We consider a C^∞-differentiable Jordan curve $\gamma : [-r, r] \to \mathbb{R}^d$, naturally parameterized such that $|\gamma'(t)| = 1$ for every $t \in [-r, r]$. The geodesic distance on the curve is then given as

$$d(\gamma(t_1), \gamma(t_2)) = \left| \int_{t_1}^{t_2} |\gamma'(t)| dt \right| = |t_1 - t_2|$$

and the length of the whole curve is $2r$. Now, for the formulation of the uncertainty principle, we adopt the notation of the previous sections. Without loss of generality we can assume that the point p at which the geodesic polar coordinates are introduced corresponds to $\gamma(0)$. Then the cut locus corresponds to the point $\gamma(r)$ and the weight function Θ satisfies $\Theta(t,\xi) = |\gamma'(\xi t)| = 1$ for all $t \in [0,r]$ and $\xi \in \{\pm 1\}$. The integration on γ can be written in the polar coordinates as

$$\int_\gamma f\,d\gamma = \sum_{\xi \in \{\pm 1\}} \int_0^r f(\gamma(\xi t))dt$$

and the Laplacian Δ_γ translates to

$$\Delta_\gamma f(\gamma(\xi t)) = \Delta_{p,t} f(\gamma(\xi t)) = \frac{d^2}{dt^2} f(\gamma(\xi t)).$$

Now, we can formulate the uncertainty principle (2.90) for the curve γ as follows.

Corollary 2.64.
If $f(\gamma(\cdot)) \in AC([-r,r])$, $f(\gamma(-r)) = f(\gamma(r))$, $\frac{d}{dt_*}f \in L^2(\gamma)$ and $\|f\|_\gamma = 1$, then the following inequality holds:

$$\frac{r^2}{\pi^2} \frac{1 - \varepsilon_p(f)^2}{\varepsilon_p(f)^2} \cdot \left\| \frac{d}{dt_*} f \right\|_\gamma^2 > \frac{1}{4}, \tag{2.168}$$

where

$$\varepsilon_p(f) = \int_{-r}^r \cos(\tfrac{\pi}{r}t)|f(\gamma(t))|^2 dt.$$

The constant $\frac{1}{4}$ on the right hand side of inequality (2.168) is optimal.

We remark that this result can also be shown in a different way. Since a smooth Jordan curve γ with length $2r$ is isometric to the circle with radius $\frac{r}{\pi}$, the uncertainty for γ can directly be deduced from the Breitenberger uncertainty principle (1.7). Moreover, this relation also shows that inequality (2.168) is optimal (see [68] for the optimality on the unit circle).

2.7. Estimates of the uncertainty principles using comparison principles

For general Riemannian manifolds with dimension $d \geq 2$, the right hand side of the inequalities (2.90), (2.110) and (2.115) is usually hard to determine. However, it is possible to simplify these terms if some further information on the curvature of the Riemannian manifold is given. The main tool in this context is Bishop's comparison theorem. A short introduction into various concepts of curvature can be found in Section A.6 of the appendix.

Theorem 2.65 (Bishop, [9], Theorem III.4.1&2).
Let $p \in M$ and assume that all sectional curvatures K of M are less than or equal to a constant κ, then

$$\kappa > 0: \quad \frac{\Theta'(t,\xi)}{\Theta(t,\xi)} > (d-1)\sqrt{\kappa}\cot(\sqrt{\kappa}t), \quad (t,\xi) \in (0, \tfrac{\pi}{\sqrt{\kappa}}) \times \mathfrak{S}_p, \tag{2.169}$$

$$\kappa = 0: \quad \frac{\Theta'(t,\xi)}{\Theta(t,\xi)} \geq (d-1)\frac{1}{t}, \quad (t,\xi) \in (0,\infty) \times \mathfrak{S}_p, \tag{2.170}$$

$$\kappa < 0: \quad \frac{\Theta'(t,\xi)}{\Theta(t,\xi)} \geq (d-1)\sqrt{-\kappa}\coth(\sqrt{-\kappa}t), \quad (t,\xi) \in (0,\infty) \times \mathfrak{S}_p, \tag{2.171}$$

and

$$\kappa > 0: \quad \Theta(t,\xi) \geq \kappa^{-\frac{d-1}{2}}\sin(\sqrt{\kappa}t)^{d-1}, \quad (t,\xi) \in (0,\tfrac{\pi}{\sqrt{\kappa}}) \times \mathfrak{S}_p, \tag{2.172}$$

$$\kappa = 0: \quad \Theta(t,\xi) \geq t^{d-1}, \quad (t,\xi) \in (0,\infty) \times \mathfrak{S}_p, \tag{2.173}$$

$$\kappa < 0: \quad \Theta(t,\xi) \geq (-\kappa)^{-\frac{d-1}{2}}\sinh(\sqrt{-\kappa}t)^{d-1}, \quad (t,\xi) \in (0,\infty) \times \mathfrak{S}_p. \tag{2.174}$$

Equality in (2.169), (2.170), (2.171) and (2.172), (2.173), (2.174) holds if and only if, for all permissible t, the ball $B(p,t) \subset M$ is isometric to a ball of radius t in the d-dimensional space M_κ of constant curvature κ ($M_\kappa = \mathbb{S}^d_{1/\sqrt{\kappa}}$, $M_0 = \mathbb{R}^d$ and $M_\kappa = \mathbb{H}^d_{1/\sqrt{-\kappa}}$ if $\kappa > 0$, $\kappa = 0$ and $\kappa < 0$, respectively).

We consider first the case when M is a compact Riemannian manifold. We assume that the Ricci curvature on M satisfies

$$\mathrm{Ric}(t\xi, t\xi) \geq \kappa_1(d-1)t^2$$

for a constant $\kappa_1 > 0$, $t \geq 0$ and all unit tangent vectors ξ in the tangent bundle TM. Then, the Bonnet-Myers Theorem [9, Theorem II.6.1] states that the distance $R(\xi)$ is bounded from above by $\frac{\pi}{\sqrt{\kappa_1}}$. Further, if we assume that all sectional curvatures on M are less than or equal to a given constant κ_2, $\kappa_2 \geq \kappa_1$, then Bishop's comparison Theorem (see equation (2.169)) states that

$$\frac{\Theta(t,\xi)'}{\Theta(t,\xi)} \geq (d-1)\sqrt{\kappa_2}\cot(\sqrt{\kappa_2}t) \tag{2.175}$$

for all $\xi \in \mathfrak{S}_p$ and $0 < t < \frac{\pi}{\sqrt{\kappa_2}}$. Moreover, the Morse-Schönberg Theorem [9, Theorem II.6.3] ensures that the distance $R(\xi)$ is bounded from below by $\frac{\pi}{\sqrt{\kappa_2}}$. Combining (2.175) and $\sqrt{\kappa_1} \leq \frac{\pi}{R(\xi)} \leq \sqrt{\kappa_2}$, we get the estimate

$$\frac{\pi}{R(\xi)}\cos(\tfrac{\pi}{R(\xi)}t) + \frac{\Theta(t,\xi)'}{\Theta(t,\xi)}\sin(\tfrac{\pi}{R(\xi)}t) \geq$$
$$\frac{\pi}{R(\xi)}\cos(\sqrt{\kappa_2}t) + (d-1)\sqrt{\kappa_2}\cot(\sqrt{\kappa_2}t)\sin(\sqrt{\kappa_1}t) \geq d\sqrt{\kappa_1}\cos(\sqrt{\kappa_2}t)$$

for all $0 < t \leq \frac{\pi}{2\sqrt{\kappa_2}}$. So, if we introduce

$$\varepsilon_p^{\kappa_2}(f) = \int_{\mathfrak{S}_p}\int_0^{\frac{\pi}{2\sqrt{\kappa_2}}}\cos(\sqrt{\kappa_2}t)|f^*(t,\xi)|^2\Theta(t,\xi)dtd\mu_p(\xi)$$

as a modified mean value, then the above assumptions ensure that

$$\varepsilon_p(f) \geq \varepsilon_p^{\kappa_2}(f)$$

holds for all functions $f \in L^2(M)$ having compact support in the ball $B(p, \frac{\pi}{2\sqrt{\kappa_2}})$ centered at p. So, modifying Theorem 2.41, we get to the following local uncertainty principle.

Corollary 2.66.
Let M be a compact Riemannian manifold $(d \geq 2)$ whose Ricci curvature fulfills

$$\text{Ric}(t\xi, t\xi) \geq \kappa_1 (d-1)t^2$$

for all unit tangent vectors $\xi \in TM$, and all of whose sectional curvatures K are less than or equal to a constant κ_2, $\kappa_2 \geq \kappa_1 > 0$. If $f \in L^2(M)$ satisfies the assumptions of Theorem 2.41 and has compact support in $B(p, \frac{\pi}{2\sqrt{\kappa_2}})$, then the following inequality holds:

$$\left(1 - \varepsilon_p^{\kappa_2}(f)^2\right) \cdot \left\| \frac{\partial}{\partial t}_* f \right\|_M^2 \geq \kappa_1 \frac{d^2}{4} \varepsilon_p^{\kappa_2}(f)^2. \tag{2.176}$$

In the case that M is a d-dimensional sphere with radius $\frac{1}{\sqrt{\kappa}}$, we have $\kappa_1 = \kappa_2 = \kappa$. Inequality (2.176) then reduces to the well known principle (see the uncertainty inequality (2.151) on the sphere $\mathbb{S}_r^d$)

$$\left(1 - \varepsilon_p(f)^2\right) \cdot \left\| \frac{\partial}{\partial t}_* f \right\|_{\mathbb{S}^d}^2 \geq \kappa \frac{d^2}{4} \varepsilon_p(f)^2.$$

Thus, the point of Corollary 2.66 is that if M is a "sphere-like" manifold where the curvature is varying only slightly around a constant κ, then the resulting uncertainty principle is also very similar to the uncertainty principle of a d-dimensional sphere with curvature κ.

Next, if Ω is a star-shaped subdomain of a Riemannian manifold M, then Bishop's comparison Theorem 2.65 implies the following modified version of Theorem 2.51.

Corollary 2.67.
Let M, $d \geq 2$, be a Riemannian manifold all of whose sectional curvatures are less than or equal to a constant κ and $\Omega \subset M$ be a compact star-shaped domain with respect to the interior point p. If $\kappa > 0$, let $Q(\xi) \leq \frac{\pi}{2\sqrt{\kappa}}$ for all $\xi \in \mathfrak{S}_p$. Further, assume that $f \in L^2(\Omega)$ fulfills the conditions of Theorem 2.51. Then, the following inequalities hold:

$$\kappa > 0: \quad \|t_* f\|_\Omega^2 \cdot \left\| \frac{\partial}{\partial t}_* f \right\|_\Omega^2 > \frac{1}{4}\left|1 + \sqrt{\kappa}(d-1)\left\|\sqrt{t\cot(\sqrt{\kappa}t)}_* f\right\|_\Omega^2\right|^2, \tag{2.177}$$

$$\kappa = 0: \quad \|t_* f\|_\Omega^2 \cdot \left\| \frac{\partial}{\partial t}_* f \right\|_\Omega^2 > \frac{d^2}{4}, \tag{2.178}$$

$$\kappa < 0: \quad \|t_* f\|_\Omega^2 \cdot \left\| \frac{\partial}{\partial t}_* f \right\|_\Omega^2 > \frac{1}{4}\left|1 + \sqrt{-\kappa}(d-1)\left\|\sqrt{t\coth(\sqrt{-\kappa}t)}_* f\right\|_\Omega^2\right|^2. \tag{2.179}$$

Finally, if E is a Riemannian manifold diffeomorphic to $\mathbb{R}^d$, we can combine the uncertainty principle (2.115) with Bishop's comparison Theorem 2.65 and get the following result.

Corollary 2.68.
Assume that E is a Riemannian manifold diffeomorphic to $\mathbb{R}^d$ all of whose sectional curvatures are less than or equal to a constant κ. Further assume that $p \in E$ and that $f \in L^2(E)$ satisfies the conditions of Theorem 2.54. Then, for $\kappa = 0$, the inequality

$$\|t_* f\|_E^2 \cdot \left\|\frac{\partial}{\partial t_*} f\right\|_E^2 \geq \frac{d^2}{4} \tag{2.180}$$

and for $\kappa < 0$, the inequality

$$\|t_* f\|_E^2 \cdot \left\|\frac{\partial}{\partial t_*} f\right\|_E^2 \geq \frac{1}{4}\left|1 + \sqrt{-\kappa}(d-1)\right\|\sqrt{t\coth(\sqrt{-\kappa}t)}_* f\|_E^2\right|^2 \tag{2.181}$$

holds. The inequalities (2.180) and (2.181) do not differ from the uncertainty inequality (2.115) if and only if E is isometric to the Euclidean space $\mathbb{R}^d$ or to the hyperbolic space $\mathrm{H}^d_{1/\sqrt{-\kappa}}$ with constant negative curvature κ, respectively.

2.8. Remarks and References

Weighted L^2-inequalities on a cylindrical domain. The weighted L^2-inequalities stated in Section 2.1 are new and can be considered as multi-dimensional extensions of the results in Section 1.4. In particular, the inequalities and the Dunkl operators of the subsections 2.1.1, 2.1.2 and 2.1.3 are multi-dimensional generalizations of the inequalities and the Dunkl operators presented in the subsections 1.4.1, 1.4.2 and 1.4.3, respectively. As in Section 1.4 also the methods used in Section 2.1 are based on the Dunkl operator approach developed in [73] and [27]. The quote in the header of page 42 is taken from: Douglas Adams, The Hitchhiker's Guide to the Galaxy, 1979.

Uncertainty principles on compact Riemannian manifolds. The uncertainty principle of Theorem 2.41, stated in this form, is an entirely novel result and can be considered as a generalization of the Breitenberger uncertainty principle (1.15) on the unit circle. Nonetheless, there exist various uncertainty principles in the literature that hold also for compact Riemannian manifolds but are based on different approaches. For a general review on various types of uncertainty principles, we refer to the survey article [20] and the book [32].
A particularly interesting uncertainty principle for compact Riemannian manifolds can be found in the recent work [56] of Martini. In [56], it is shown that for all $\alpha, \beta > 0$ and $f \in L^2(M)$ with null mean value the following inequality holds:

$$\|f\|_M \leq C_{\alpha,\beta}\|t_* f\|_M^{\frac{\beta}{\alpha+\beta}} \cdot \|(-\Delta_M)^{\frac{\beta}{2}} f\|_M^{\frac{\alpha}{\alpha+\beta}}.$$

This inequality is a special case of a more general theory treating uncertainty principles on abstract measure spaces (see also [11] and [72]). The proof of this inequality is mainly based on the spectral theorem and on estimates involving the heat semigroup generated by the Laplace-Beltrami operator Δ_M. In contrast to the uncertainty inequality (2.94), the constant $C_{\alpha,\beta}$ in the above inequality is not explicitly known.

Another interesting uncertainty principle for compact manifolds, a generalization of the uncertainty principle of Hardy that is based on the eigenfunction expansion of the Laplace-Beltrami operator Δ_M, can be found in the article [65].

Sections 2.3 and 2.4. Theorem 2.54 in Section 2.4 can be seen as a generalization of the classical Heisenberg-Pauli-Weyl inequality (1.9). A variety of uncertainty principles that are related to the uncertainty inequality (2.115) can be found in the literature, in particular, in [20], [32] and [56]. However, Theorem 2.54 in this form is new and particularly interesting since the included uncertainty inequality is sharp and the underlying proof is based on an operator theoretic approach that provides a position and a frequency variance of a function f.

Theorem 2.51 in Section 2.3 is also a new result and has strong relations to the uncertainty principles in Sections 2.2 and 2.4. In particular, if Ω is a compact star-shaped subset of a manifold E diffeomorphic to $\mathbb{R}^d$, then Theorem 2.51 is an immediate consequence of Theorem 2.54.

Asymptotic sharpness. The usage of a Gaussian-type function instead of the heat kernel to prove the asymptotic sharpness of the uncertainty principles in the Propositions 2.58 and 2.59 is novel in this thesis. In prior works, Fourier techniques and the heat kernel were used to prove the asymptotic sharpness of the uncertainty principles for ultraspherical expansions [73], for Jacobi expansions [54] and on the unit circle [68].

Uncertainty principles on the unit sphere. In the literature, there exist several uncertainty principles on the unit sphere $\mathbb{S}^d$ that are very similar to the one in Corollary 2.60. In the first place, we mention the article [73] in which Corollary 2.60 was proven for radial functions on the unit sphere $\mathbb{S}^d$, i.e. functions that have an expansion in terms of ultraspherical polynomials.

Other works treating uncertainty principles on the sphere attained similar results as in Corollary 2.60, but worked with slightly different techniques. In [62], the Laplace-Beltrami operator $\Delta_{\mathbb{S}^2}$ was used to define a frequency variance for functions on $\mathbb{S}^2$. For the proof of the uncertainty inequality, a vector valued differential operator was introduced to split the operator $\Delta_{\mathbb{S}^2}$. Similar results using the same technique as in [62] were also obtained in [21, Section 5.5] and [22].

Later on, also in [26] and [27] vector valued differential operators were used to prove an uncertainty principle on $\mathbb{S}^d$ that is very similar to the uncertainty in Corollary 2.60.

Uncertainty principles on projective spaces. Corollary 2.61 is a novel result but strongly related to uncertainty principles for Jacobi expansions. In fact, since the radial square integrable functions on a projective space have an expansion in terms of Jacobi polynomials, Corollary 2.61 restricted to radial functions corresponds exactly to the uncertainty principle proven in [54]. The technical details of the projective spaces are primarily taken from the books [3], [4] and the article [35].

Uncertainty principles on hyperbolic spaces. The uncertainty principle on hyperbolic spaces formulated in Corollary 2.63 is novel. Another interesting local uncertainty principle for hyperbolic spaces can be found in [82].

Estimates of the uncertainty principles using comparison principles. The usage of comparison theorems, in particular Bishop's theorem, to estimate the uncertainty principles explicitly in terms of the curvature of the Riemannian manifold is novel in this thesis. All the comparison theorems used in Section 2.7, including Bishop's Theorem, the Morse-Schönberg Theorem and the Bonnet-Myers Theorem used are taken from [9].

3

Optimally space localized
polynomials

In this final chapter, we are going to study how the uncertainty principles of the previous
chapters can be used to find well-localized polynomials and to analyze the space-frequency
behavior of certain families of polynomials. In principle, we will consider two particular
settings: Jacobi expansions on the interval $[0, \pi]$ and spherical polynomials on compact
two-point homogeneous spaces.

In the first section, we will study the uncertainty inequality (1.72) for Jacobi expansions
with regard to polynomial subspaces of $L^2([0, \pi], w_{\alpha\beta})$. In particular, in Theorem 3.6
and in Corollary 3.10, we will give representations of those polynomials $\mathcal{P}_n^{(\alpha,\beta)}$ that are
optimally localized at the left hand boundary of the interval $[0, \pi]$ with respect to the
mean value $\varepsilon_{\alpha\beta}$. We will show (Proposition 3.7) that the position variance $\mathrm{var}_S^{\alpha\beta}$ of the
uncertainty principle gets minimal for the polynomials $\mathcal{P}_n^{(\alpha,\beta)}$. Moreover, in Theorem 3.14,
we will prove that the uncertainty product $\mathrm{var}_S^{\alpha\beta}(\mathcal{P}_n^{(\alpha,\beta)}) \cdot \mathrm{var}_F^{\alpha\beta}(\mathcal{P}_n^{(\alpha,\beta)})$ of the optimally
space localized polynomials $\mathcal{P}_n^{(\alpha,\beta)}$ is uniformly bounded but does in general not tend to
the optimal constant as $n \to \infty$. Finally, we will analyze the space-frequency behavior
of two further well-known families of polynomials, the Christoffel-Darboux kernels and
the de La Vallée Poussin kernels. As a consequence of Theorem 3.15, we will see that the
uncertainty product for the Christoffel-Darboux kernel $\tilde{K}_n^{(\alpha,\beta)}$ tends linearly to infinity
as $n \to \infty$, whereas the uncertainty product of the de La Vallée Poussin kernel $\tilde{V}_n$ in
Theorem 3.16 tends to the optimal constant of the uncertainty principle.

The second section includes an intermediate result on the monotonicity of extremal zeros
of Jacobi and associated Jacobi polynomials when certain parameters are altered. This

auxiliary result enables us to carry the optimality results of Theorem 3.6 over to the setting of spherical polynomials on a compact two-point homogeneous space M.

In the last section, we will then investigate the uncertainty product on compact two-point homogeneous spaces M, i.e., the spheres $\mathbb{S}_r^d$ and the projective spaces $\mathbb{RP}_r^d$, $\mathbb{CP}_r^d$, $\mathbb{HP}_r^d$ and $\mathbb{C}a_r$, in relation with certain spaces of spherical polynomials on M. Similar as in Section 3.1, we will give in Theorem 3.28 and Corollary 3.29 explicit formulas for those polynomials $\mathcal{P}_n^M$ that are optimally localized at a point $p \in M$ with respect to the mean value ε_p^M and that minimize the position variance $\mathrm{var}_{\mathbb{S},p}^M$ of the uncertainty principle.

3.1. Optimally space localized polynomials for Jacobi expansions

We start out by introducing particular polynomial subspaces of the Hilbert space $L^2([0,\pi], w_{\alpha\beta})$. As in Section 1.5.1, the weight function $w_{\alpha\beta}$ given by

$$w_{\alpha\beta}(t) = 2^{\alpha+\beta+1} \sin^{2\alpha+1}\left(\frac{t}{2}\right) \cos^{2\beta+1}\left(\frac{t}{2}\right)$$

denotes for $\alpha, \beta > -1$ the Jacobi weight on the interval $[0,\pi]$ and $P_l^{(\alpha,\beta)}(\cos t)$ the Jacobi polynomial of order l. Further, we define by

$$p_l^{(\alpha,\beta)}(\cos t) := \frac{P_l^{(\alpha,\beta)}(\cos t)}{\|P_l^{(\alpha,\beta)}\|_{w_{\alpha\beta}}} \tag{3.1}$$

the respective orthonormal Jacobi polynomial on $[0,\pi]$.

Definition 3.1. As subspaces of $L^2([0,\pi], w_{\alpha\beta})$, we consider the following three polynomial spaces:

(1) The space spanned by the polynomials $p_l^{(\alpha,\beta)}$, $l \leq n$:

$$\Pi_n^{(\alpha,\beta)} := \left\{ P : \ P(t) = \sum_{l=0}^{n} c_l p_l^{(\alpha,\beta)}(\cos t), \ c_0, \ldots, c_n \in \mathbb{C} \right\}. \tag{3.2}$$

(2) The space spanned by the polynomials $p_l^{(\alpha,\beta)}$, $m \leq l \leq n$:

$$\Pi_{m,n}^{(\alpha,\beta)} := \left\{ P : \ P(t) = \sum_{l=m}^{n} c_l p_l^{(\alpha,\beta)}(\cos t), \ c_m, \ldots, c_n \in \mathbb{C} \right\}. \tag{3.3}$$

(3) The space spanned by a polynomial $\mathcal{R}(t) = p_m^{(\alpha,\beta)}(\cos t) + \sum_{l=0}^{m-1} e_l p_l^{(\alpha,\beta)}(\cos t)$ of degree

m and the polynomials $p_l^{(\alpha,\beta)}$, $m + 1 \leq l \leq n$:

$$\Pi_{\mathcal{R},n}^{(\alpha,\beta)} := \left\{ P : \ P(t) = c_m \mathcal{R}(t) + \sum_{l=m+1}^{n} c_l p_l^{(\alpha,\beta)}(\cos t), \ c_m, \ldots, c_n \in \mathbb{C} \right\}. \qquad (3.4)$$

Further, we define the unit spheres of the spaces $\Pi_n^{(\alpha,\beta)}$, $\Pi_{m,n}^{(\alpha,\beta)}$ and $\Pi_{\mathcal{R},n}^{(\alpha,\beta)}$ as

$$\mathbb{S}_n^{(\alpha,\beta)} := \left\{ P \in \Pi_n^{(\alpha,\beta)} : \ \|P\|_{w_{\alpha\beta}} = 1 \right\},$$
$$\mathbb{S}_{m,n}^{(\alpha,\beta)} := \left\{ P \in \Pi_{m,n}^{(\alpha,\beta)} : \ \|P\|_{w_{\alpha\beta}} = 1 \right\},$$
$$\mathbb{S}_{\mathcal{R},n}^{(\alpha,\beta)} := \left\{ P \in \Pi_{\mathcal{R},n}^{(\alpha,\beta)} : \ \|P\|_{w_{\alpha\beta}} = 1 \right\}.$$

Remark 3.2. Clearly, $\Pi_{m,n}^{(\alpha,\beta)} \subset \Pi_n^{(\alpha,\beta)}$ and $\Pi_{\mathcal{R},n}^{(\alpha,\beta)} \subset \Pi_n^{(\alpha,\beta)}$. In the literature, the spaces $\Pi_{m,n}^{(\alpha,\beta)}$ are sometimes called wavelet spaces and considered in a more general theory on polynomial wavelets and polynomial frames, see for instance [58] and the references therein. The standardization $e_m = 1$ for the highest expansion coefficient of the polynomial $\mathcal{R}$ causes no loss of generality and is a useful convention for the upcoming calculations. The polynomials in the spaces $\Pi_{\mathcal{R},n}^{(\alpha,\beta)}$ play an important role in the theory of polynomial approximation. Hereby, a usual choice for the polynomial $\mathcal{R}$ is the Christoffel-Darboux kernel $K_m^{(\alpha,\beta)}(t)$ of order m given by

$$K_m^{(\alpha,\beta)}(t) := \sum_{l=0}^{m} p_l^{(\alpha,\beta)}(1) p_l^{(\alpha,\beta)}(\cos t).$$

As contemporary references on this topic we refer to [17] and [59].

The first goal of this section is to study the localization of the polynomials in the spaces $\Pi_n^{(\alpha,\beta)}$, $\Pi_{m,n}^{(\alpha,\beta)}$ and $\Pi_{\mathcal{R},n}^{(\alpha,\beta)}$ at the left hand boundary of the interval $[0, \pi]$ and to determine those polynomials that are in some sense best localized. As an analyzing tool for the localization of a function $f \in L^2([0, \pi], w_{\alpha\beta})$ at the point $t = 0$, we consider the mean value

$$\varepsilon_{\alpha\beta}(f) = \int_0^{\pi} \cos t \, |f(t)|^2 w_{\alpha\beta}(t) dt, \qquad (3.5)$$

as defined in (1.71). If $\|f\|_{w_{\alpha\beta}} = 1$, then $-1 < \varepsilon_{\alpha\beta}(f) < 1$, and the more the mass of the L^2-density f is concentrated at the boundary point $t = 0$, the closer the value $\varepsilon_{\alpha\beta}(f)$ gets to 1. Therefore, the value $\varepsilon_{\alpha\beta}(f)$ can be interpreted as a measure on how well the function f is localized at the left hand boundary of the interval $[0, \pi]$. We say that f is localized at $t = 0$ if the value $\varepsilon_{\alpha\beta}(f)$ approaches 1.

Now, our aim is to find those elements of the polynomial spaces $\Pi_n^{(\alpha,\beta)}$, $\Pi_{m,n}^{(\alpha,\beta)}$ and $\Pi_{\mathcal{R},n}^{(\alpha,\beta)}$ that are optimally localized at the boundary point $t = 0$. In particular, we want to solve the following optimization problems:

$$\mathcal{P}_n^{(\alpha,\beta)} = \arg \max_{P \in \mathbb{S}_n^{(\alpha,\beta)}} \varepsilon_{\alpha\beta}(P), \tag{3.6}$$

$$\mathcal{P}_{m,n}^{(\alpha,\beta)} = \arg \max_{P \in \mathbb{S}_{m,n}^{(\alpha,\beta)}} \varepsilon_{\alpha\beta}(P), \tag{3.7}$$

$$\mathcal{P}_{\mathcal{R},n}^{(\alpha,\beta)} = \arg \max_{P \in \mathbb{S}_{\mathcal{R},n}^{(\alpha,\beta)}} \varepsilon_{\alpha\beta}(P). \tag{3.8}$$

Since the linear spaces $\Pi_n^{(\alpha,\beta)}$, $\Pi_{m,n}^{(\alpha,\beta)}$ and $\Pi_{\mathcal{R},n}^{(\alpha,\beta)}$ are finite-dimensional, the unit spheres $\mathbb{S}_n^{(\alpha,\beta)}$, $\mathbb{S}_{m,n}^{(\alpha,\beta)}$ and $\mathbb{S}_{\mathcal{R},n}^{(\alpha,\beta)}$ are compact subsets and the functional $\varepsilon_{\alpha\beta}$ is bounded and continuous on the respective polynomial space. Hence, it is guaranteed that solutions of the optimization problems (3.6), (3.7) and (3.8) exist.

Remark 3.3. Instead of searching for the polynomials $\mathcal{P}_n^{(\alpha,\beta)}$, $\mathcal{P}_{m,n}^{(\alpha,\beta)}$ and $\mathcal{P}_{\mathcal{R},n}^{(\alpha,\beta)}$ that are optimally localized at $t = 0$ and that maximize the mean value $\varepsilon_{\alpha\beta}$, we could also search for the polynomials that are optimally localized at the right hand boundary of the interval $[0, \pi]$ and that minimize the mean value $\varepsilon_{\alpha\beta}$. Since the weight function $w_{\alpha\beta}$ and the Jacobi polynomial $p_n^{(\alpha,\beta)}$ satisfy $w_{\alpha\beta}(t) = w_{\beta\alpha}(\pi - t)$ and $p_n^{(\alpha,\beta)}(\cos t) = (-1)^n p_n^{(\beta,\alpha)}(\cos(\pi - t))$, minimizing $\varepsilon_{\alpha\beta}(P)$ with respect to polynomials in $\mathbb{S}_n^{(\alpha,\beta)}$, $\mathbb{S}_{m,n}^{(\alpha,\beta)}$ and $\mathbb{S}_{\mathcal{R},n}^{(\alpha,\beta)}$ yields the same as maximizing $\varepsilon_{\beta\alpha}(P)$ with respect to polynomials in $\mathbb{S}_n^{(\beta,\alpha)}$, $\mathbb{S}_{m,n}^{(\beta,\alpha)}$ and $\mathbb{S}_{\mathcal{R},n}^{(\beta,\alpha)}$. Because of this symmetric relation, it is entirely sufficient to consider only the maximization problems (3.6), (3.7) and (3.8).

In order to describe the optimal polynomials, we need the notion of associated and of scaled co-recursive associated polynomials. First of all, we know that the orthonormal Jacobi polynomials $p_l^{(\alpha,\beta)}$ satisfy the three-term recurrence relation (see [25, Table 1.1])

$$b_{l+1}p_{l+1}^{(\alpha,\beta)}(x) = (x - a_l)p_l^{(\alpha,\beta)}(x) - b_l p_l^{(\alpha,\beta)}(x), \quad l = 0,1,2,3,\ldots \tag{3.9}$$

$$p_{-1}^{(\alpha,\beta)}(x) = 0, \qquad p_0^{(\alpha,\beta)}(x) = \frac{1}{b_0},$$

for $x = \cos t$ in the interval $[-1, 1]$ and

$$a_l = \frac{\beta^2 - \alpha^2}{(2l + \alpha + \beta)(2l + \alpha + \beta + 2)}, \quad l = 0,1,2,\ldots \tag{3.10}$$

$$b_l = \left(\frac{4l(l+\alpha)(l+\beta)(l+\alpha+\beta)}{(2l+\alpha+\beta)^2(2l+\alpha+\beta+1)(2l+\alpha+\beta-1)} \right)^{\frac{1}{2}}, \quad l = 1,2,3,\ldots \tag{3.11}$$

$$b_0 = \left(\int_0^\pi w_{\alpha\beta}dt \right)^{\frac{1}{2}} = \left(2^{\alpha+\beta+1}\frac{\Gamma(\alpha+1)\Gamma(\beta+1)}{\Gamma(\alpha+\beta+2)} \right)^{\frac{1}{2}}.$$

Then, the associated and the scaled co-recursive associated Jacobi polynomials are defined as follows:

Definition 3.4. For $c \geq 0$ (if $c > 0$, assume that $\alpha + \beta \neq -2c$), we define the associated Jacobi polynomials $p_l^{(\alpha,\beta)}(x,c)$ on the interval $[-1,1]$ by the shifted recurrence relation

$$b_{c+l+1}\, p_{l+1}^{(\alpha,\beta)}(x,c) = (x - a_{c+l})\, p_l^{(\alpha,\beta)}(x,c) - b_{c+l}\, p_{l-1}^{(\alpha,\beta)}(x,c), \quad l = 0, 1, 2, \ldots, \tag{3.12}$$

$$p_{-1}^{(\alpha,\beta)}(x,c) = 0, \qquad p_0^{(\alpha,\beta)}(x,c) = 1.$$

Further, for $\gamma \in \mathbb{R}$ and $\delta \geq 0$, we define the scaled co-recursive associated Jacobi polynomials $p_l^{(\alpha,\beta)}(x,c,\gamma,\delta)$ on $[-1,1]$ by the three-term recurrence relation

$$b_{c+l+1}\, p_{l+1}^{(\alpha,\beta)}(x,c,\gamma,\delta) = (x - a_{c+l})\, p_l^{(\alpha,\beta)}(x,c,\gamma,\delta) - b_{c+l}\, p_{l-1}^{(\alpha,\beta)}(x,c,\gamma,\delta),$$
$$l = 1, 2, 3, 4 \ldots, \tag{3.13}$$

$$p_0^{(\alpha,\beta)}(x,c,\gamma,\delta) = 1, \quad p_1^{(\alpha,\beta)}(x,c,\gamma,\delta) = \frac{\delta x - a_c - \gamma}{\beta_{c+1}}.$$

The three-term recurrence relation of the co-recursive associated Jacobi polynomials $p_{l+1}^{(\alpha,\beta)}(x,c,\gamma,\delta)$ corresponds to the three-term recurrence relation of the associated Jacobi polynomials except for the formula of the initial polynomial $p_1^{(\alpha,\beta)}(x,c,\gamma,\delta)$. For $c = 0$, $\gamma = 0$ and $\delta = 1$, we have the identities $p_l^{(\alpha,\beta)}(x,0) = p_l^{(\alpha,\beta)}(x,0,0,1) = b_0\, p_l^{(\alpha,\beta)}(x)$. For $m \in \mathbb{N}$, the associated polynomials $p_l^{(\alpha,\beta)}(x,m)$ and $p_l^{(\alpha,\beta)}(x,m,\gamma,\delta)$ can be described with help of the symmetric Jacobi matrix $\mathbf{J}_n^m$, $0 \leq m \leq n$, defined by

$$\mathbf{J}_n^m = \begin{pmatrix} a_m & b_{m+1} & 0 & 0 & \cdots & 0 \\ b_{m+1} & a_{m+1} & b_{m+2} & 0 & \cdots & 0 \\ 0 & b_{m+2} & a_{m+2} & b_{m+3} & \ddots & \vdots \\ \vdots & \ddots & \ddots & \ddots & \ddots & 0 \\ 0 & \cdots & 0 & b_{n-2} & a_{n-1} & b_{n-1} \\ 0 & \cdots & & 0 & b_{n-1} & a_n \end{pmatrix}. \tag{3.14}$$

If $m = 0$, we write $\mathbf{J}_n$ instead of $\mathbf{J}_n^0$. Then, in view of the three-term recurrence formulas (3.12) and (3.13), the polynomials $p_l^{(\alpha,\beta)}(x,m)$ and $p_l^{(\alpha,\beta)}(x,m,\gamma,\delta)$, $l \geq 1$, can be written as (cf. [42, Theorem 2.2.4])

$$p_l^{(\alpha,\beta)}(x,m) = \det(x\mathbf{1}_l - \mathbf{J}_{m+l-1}^m), \tag{3.15}$$

and

$$p_l^{(\alpha,\beta)}(x,m,\gamma,\delta) = \det\left(x \begin{pmatrix} \delta & 0 \\ 0 & \mathbf{1}_{l-1} \end{pmatrix} - \mathbf{J}_{m+l-1}^m - \begin{pmatrix} \gamma & 0 \\ 0 & \mathbf{0}_{l-1} \end{pmatrix} \right), \tag{3.16}$$

where $\mathbf{1}_{l-1}$ denotes the $(l-1)$-dimensional identity matrix and $\mathbf{0}_{l-1}$ the $(l-1)$-dimensional zero matrix.

Next, we give a characterization of the mean value $\varepsilon_{\alpha\beta}(P)$ in terms of the expansion coefficients c_l of the polynomial $P = \sum_{l=n}^m c_l p_l^{(\alpha,\beta)}$.

Lemma 3.5.

For the polynomial $P(t) = \sum\limits_{l=0}^{n} c_l p_l^{(\alpha,\beta)}(\cos t)$, we have

$$\varepsilon_{\alpha\beta}(P) = \mathbf{c}^H \mathbf{J}_n \mathbf{c}, \qquad\qquad\qquad\qquad \text{if } P \in \Pi_n^{(\alpha,\beta)},$$

$$\varepsilon_{\alpha\beta}(P) = \tilde{\mathbf{c}}^H \mathbf{J}_n^m \tilde{\mathbf{c}}, \qquad\qquad\qquad\qquad \text{if } P \in \Pi_{m,n}^{(\alpha,\beta)},$$

$$\varepsilon_{\alpha\beta}(P) = \tilde{\mathbf{c}}^H \mathbf{J}_n^m \tilde{\mathbf{c}} + (\varepsilon_{\alpha\beta}(\mathcal{R}) - a_m)|c_m|^2, \qquad \text{if } P \in \Pi_{\mathcal{R},n}^{(\alpha,\beta)},$$

with the coefficient vectors $\mathbf{c} = (c_0, \ldots, c_n)^T$ and $\tilde{\mathbf{c}} = (c_m, \ldots, c_n)^T$.

Proof. Using the three-term recurrence formula (3.9) and the orthonormality relation of the Jacobi polynomials $p_l^{(\alpha,\beta)}$, we get for $P \in \Pi_n^{(\alpha,\beta)}$

$$\varepsilon_{\alpha\beta}(P) = \int_0^\pi \cos t \left| \sum_{l=0}^{n} c_l p_l^{(\alpha,\beta)}(\cos t) \right|^2 w_{\alpha\beta}(t)dt$$

$$= \int_0^\pi \left(\sum_{l=0}^{n} c_l \cos t\, p_l^{(\alpha,\beta)}(\cos t) \right) \overline{\left(\sum_{l=0}^{n} c_l p_l^{(\alpha,\beta)}(\cos t) \right)} w_{\alpha\beta}(t)dt$$

$$= \int_0^\pi \left(\sum_{l=0}^{n} c_l \left(b_{l+1} p_{l+1}^{(\alpha,\beta)}(\cos t) + a_l p_l^{(\alpha,\beta)}(\cos t) + b_l p_{l-1}^{(\alpha,\beta)}(\cos t) \right) \right)$$

$$\times \overline{\left(\sum_{l=0}^{n} c_l p_l^{(\alpha,\beta)}(\cos t) \right)} w_{\alpha\beta}(t)dt$$

$$= \sum_{l=0}^{n} a_l |c_l|^2 + \sum_{l=0}^{n-1} \left(b_{l+1} c_l \bar{c}_{l+1} + b_{l+1} \bar{c}_l c_{l+1} \right) = \mathbf{c}^H \mathbf{J}_n \mathbf{c}.$$

If $c_0 = \ldots = c_{m-1} = 0$, we get the assertion for polynomials P in the space $\Pi_{m,n}^{(\alpha,\beta)}$. If $P \in \Pi_{\mathcal{R},n}^{(\alpha,\beta)}$, then P has the representation

$$P(t) = c_m \left(p_m^{(\alpha,\beta)}(\cos t) + \sum_{l=0}^{m-1} e_l p_l^{(\alpha,\beta)}(\cos t) \right) + \sum_{l=m+1}^{n} c_l p_l^{(\alpha,\beta)}(\cos t),$$

where the polynomial $\mathcal{R}$ is given by $\mathcal{R}(t) = p_m^{(\alpha,\beta)}(\cos t) + \sum_{l=0}^{m-1} e_l p_l^{(\alpha,\beta)}(\cos t)$. Inserting this representation in the upper formula for $\varepsilon_{\alpha\beta}(P)$ yields the identity $\varepsilon_{\alpha\beta}(P) = (\varepsilon_{\alpha\beta}(\mathcal{R}) - a_m)|c_m|^2 + \tilde{\mathbf{c}}^H \mathbf{J}_n^m \tilde{\mathbf{c}}$. □

Using the characterization of $\varepsilon_{\alpha\beta}(P)$ in Lemma 3.5, we proceed to the solution of the optimization problems (3.6), (3.7) and (3.8).

Theorem 3.6.

The solutions of the optimization problems (3.6), (3.7) and (3.8) are given by

$$\mathcal{P}_n^{(\alpha,\beta)}(t) = \kappa_1 \sum_{l=0}^{n} p_l^{(\alpha,\beta)}(\lambda_{n+1})\, p_l^{(\alpha,\beta)}(\cos t), \qquad\qquad (3.17)$$

$$\mathcal{P}_{m,n}^{(\alpha,\beta)}(t) = \kappa_2 \sum_{l=m}^{n} p_{l-m}^{(\alpha,\beta)}(\lambda_{n-m+1}^{m}, m) \, p_l^{(\alpha,\beta)}(\cos t), \qquad (3.18)$$

$$\mathcal{P}_{\mathcal{R},n}^{(\alpha,\beta)}(t) = \kappa_3 \left(\mathcal{R}(t) + \sum_{l=m+1}^{n} p_{l-m}^{(\alpha,\beta)}(\lambda_{n-m+1}^{\mathcal{R}}, m, \gamma_{\mathcal{R}}, \delta_{\mathcal{R}}) p_l^{(\alpha,\beta)}(\cos t) \right), \qquad (3.19)$$

*where $p_l^{(\alpha,\beta)}(x,m)$ and $p_l^{(\alpha,\beta)}(x,m,\gamma_{\mathcal{R}},\delta_{\mathcal{R}})$ denote the associated and the scaled co-recursive associated Jacobi polynomials as given in Definition 3.4 with the shift term $\gamma_{\mathcal{R}} := \varepsilon_{\alpha\beta}(\mathcal{R}) - a_m$ and the scaling factor $\delta_{\mathcal{R}} := \|\mathcal{R}\|_{w_{\alpha\beta}}^2$.
The values λ_{n+1}, λ_{n-m+1}^{m} and $\lambda_{n-m+1}^{\mathcal{R}}$ denote the largest zero of the polynomials $p_{n+1}^{(\alpha,\beta)}(x)$, $p_{n-m+1}^{(\alpha,\beta)}(x,m)$ and $p_{n-m+1}^{(\alpha,\beta)}(x,m,\gamma_{\mathcal{R}},\delta_{\mathcal{R}})$ in the interval $[-1,1]$, respectively. The constants κ_1, κ_2 and κ_3 are chosen such that the optimal polynomials lie in the respective unit sphere and are uniquely determined up to multiplication with a complex scalar of absolute value one. The maximal value of $\varepsilon_{\alpha\beta}$ in the respective polynomial space is given by*

$$M_n^{(\alpha,\beta)} := \max_{P \in \mathbb{S}_n^{(\alpha,\beta)}} \varepsilon_{\alpha\beta}(P) = \lambda_{n+1},$$

$$M_{m,n}^{(\alpha,\beta)} := \max_{P \in \mathbb{S}_{m,n}^{(\alpha,\beta)}} \varepsilon_{\alpha\beta}(P) = \lambda_{n-m+1}^{m},$$

$$M_{\mathcal{R},n}^{(\alpha,\beta)} := \max_{P \in \mathbb{S}_{\mathcal{R},n}^{(\alpha,\beta)}} \varepsilon_{\alpha\beta}(P) = \lambda_{n-m+1}^{\mathcal{R}}.$$

Proof. We start out by determining the optimal solution $\mathcal{P}_{m,n}^{(\alpha,\beta)}$ for the optimization problem (3.7). The formula for the the optimal polynomial $\mathcal{P}_n^{(\alpha,\beta)}$ follows then as a special case if we set $m = 0$. First of all, Lemma 3.5 states that the mean value $\varepsilon_{\alpha\beta}(P)$ of a polynomial $P(t) = \sum_{l=m}^{n} c_l p_l^{(\alpha,\beta)}(\cos t)$ can be written as $\varepsilon_{\alpha\beta}(P) = \tilde{\mathbf{c}}^H \mathbf{J}_n^m \tilde{\mathbf{c}}$ with the coefficient vector $\tilde{\mathbf{c}} = (c_m, \cdots, c_n)^T$. Thus, maximizing $\varepsilon_{\alpha\beta}(P)$ with respect to a normed polynomial $P \in \mathbb{S}_{m,n}^{(\alpha,\beta)}$ is equivalent to maximize the quadratic functional $\tilde{\mathbf{c}}^H \mathbf{J}_n^m \tilde{\mathbf{c}}$ subject to $|\tilde{\mathbf{c}}|^2 = c_m^2 + c_{m+1}^2 + \cdots + c_n^2 = 1$. If λ_{n-m+1}^{m} denotes the largest eigenvalue of the symmetric Jacobi matrix $\mathbf{J}_n^m$, we have

$$\tilde{\mathbf{c}}^H \mathbf{J}_n^m \tilde{\mathbf{c}} \leq \lambda_{n-m+1}^{m} |\tilde{\mathbf{c}}|^2 \qquad (3.20)$$

and equality is attained for the eigenvectors corresponding to λ_{n-m+1}^{m}. Now, the largest eigenvalue of the Jacobi matrix $\mathbf{J}_n^m$ corresponds exactly with the largest zero of the associated Jacobi polynomial $p_{n-m+1}^{(\alpha,\beta)}(x,m)$ (cf. [25, Theorem 1.31]). Using the recursion formula (3.12) of the associated Jacobi polynomials $p_l^{(\alpha,\beta)}(x,m)$ with $c_m = 1$ the eigenvalue equation $\mathbf{J}_n^m \tilde{\mathbf{c}} = \lambda_{n-m+1}^{m} \tilde{\mathbf{c}}$ yields

$$c_l = p_{l-m}^{(\alpha,\beta)}(\lambda_{n-m+1}^{m}, m), \quad l = m, \ldots n.$$

Finally, we have to normalize the coefficients c_l, $m \leq l \leq n$, such that $|\tilde{\mathbf{c}}|^2 = 1$. This is done by the absolute value of the constant κ_2. The uniqueness (up to a complex scalar with absolute value 1) of the optimal polynomial $\mathcal{P}_n^{(\alpha,\beta)}$ follows from the fact that the largest zero of $p_{n-m+1}^{(\alpha,\beta)}(x,m)$ is simple (see [10, Theorem 5.3]). The formula for $M_{m,n}^{(\alpha,\beta)}$ follows directly from the estimate in (3.20).

We consider now the third polynomial space $\Pi_{\mathcal{R},n}^{(\alpha,\beta)}$. Lemma 3.5 states that in this case the mean value $\varepsilon_{\alpha\beta}(P)$ of $P(t) = c_m \mathcal{R}(t) + \sum_{l=m+1}^{n} c_l p_l^{(\alpha,\beta)}(\cos t)$ can be written as $\varepsilon_{\alpha\beta}(P) = \tilde{\mathbf{c}}^H \mathbf{J}_n^m \tilde{\mathbf{c}} + (\varepsilon_{\alpha\beta}(\mathcal{R}) - a_m)|c_m|^2$, with the coefficient vector $\tilde{\mathbf{c}} = (c_m, \cdots, c_n)^T$. Maximizing $\varepsilon_{\alpha\beta}(P)$ with respect to a polynomial $P \in \mathbb{S}_{\mathcal{R},n}^{(\alpha,\beta)}$ is therefore equivalent to maximize the quadratic functional $\tilde{\mathbf{c}}^H \mathbf{J}_n^m \tilde{\mathbf{c}} + (\varepsilon_{\alpha\beta}(\mathcal{R}) - a_m)|c_m|^2$ subject to $(\|\mathcal{R}\|_{w_{\alpha\beta}}^2 - 1)|c_m|^2 + |\tilde{\mathbf{c}}|^2 = 1$. Using a Lagrange multiplier λ and differentiating the Lagrange function, we obtain the identity

$$\mathbf{J}_n^m \tilde{\mathbf{c}} + \gamma_{\mathcal{R}}(c_m, 0, \cdots, 0)^T = \lambda\Big(\delta_{\mathcal{R}} c_m, c_{m+1}, \cdots, c_n\Big)^T$$

as a necessary condition for the maximum, where $\gamma_{\mathcal{R}} = \varepsilon_{\alpha\beta}(\mathcal{R}) - a_m$ and $\delta_{\mathcal{R}} = \|\mathcal{R}\|_{w_{\alpha\beta}}^2$. By the equation (3.16), this system of equations is related to the three-term recursion formula (3.13) of the scaled co-recursive associated polynomials $p_l^{(\alpha,\beta)}(x, m, \gamma_{\mathcal{R}}, \delta_{\mathcal{R}})$. In particular, the value λ corresponds to a root of $p_{n-m+1}^{(\alpha,\beta)}(x, m, \gamma_{\mathcal{R}}, \delta_{\mathcal{R}})$. Moreover, the maximum of $\tilde{\mathbf{c}}^H \mathbf{J}_n^m \tilde{\mathbf{c}} + \gamma_{\mathcal{R}}|c_m|^2$ is attained for the largest root $\lambda = \lambda_{n-m+1}^{\mathcal{R}}$ of $p_{n-m+1}^{(\alpha,\beta)}(x, m, \gamma_{\mathcal{R}}, \delta_{\mathcal{R}})$ and the corresponding eigenvector

$$\tilde{\mathbf{c}} = \kappa_3\Big(1, p_1^{(\alpha,\beta)}(\lambda_{n-m+1}^{\mathcal{R}}, m, \gamma_{\mathcal{R}}, \delta_{\mathcal{R}}), \ldots, p_{n-m}^{(\alpha,\beta)}(\lambda_{n-m+1}^{\mathcal{R}}, m, \gamma_{\mathcal{R}}, \delta_{\mathcal{R}})\Big)^T,$$

where the constant κ_3 is chosen such that the condition $(\delta_{\mathcal{R}} - 1)|c_m|^2 + |\tilde{\mathbf{c}}|^2 = 1$ is satisfied. The uniqueness of the polynomial $\mathcal{P}_{\mathcal{R},n}^{(\alpha,\beta)}$ (up to a complex scalar of absolute value one) follows from the simplicity of the largest root $\lambda_{n-m+1}^{\mathcal{R}}$ of the polynomials $p_l^{(\alpha,\beta)}(x, m, \gamma_{\mathcal{R}}, \delta_{\mathcal{R}})$ (see [10, Theorem 5.3]). From the above argumentation it is also clear that the maximal value $M_{\mathcal{R},n}^{(\alpha,\beta)}$ is precisely the largest eigenvalue $\lambda_{n-m+1}^{\mathcal{R}}$. $\qquad \square$

In Corollary 1.34, the uncertainty principle for functions $f \in L^2([0,\pi], w_{\alpha\beta})$ was formulated in terms of the following position variance:

$$\text{var}_S^{\alpha\beta}(f) = \frac{1 - \varepsilon_{\alpha\beta}(f)^2}{\Big(\frac{\alpha-\beta}{\alpha+\beta+2} + \varepsilon_{\alpha\beta}(f)\Big)^2}. \tag{3.21}$$

As the mean value $\varepsilon_{\alpha\beta}(f)$, also the position variance $\text{var}_S^{\alpha\beta}(f)$ measures the localization of the function f at the boundary point $t = 0$ of the interval $[0,\pi]$. In fact, if we define the subsets

$$\mathcal{L}_n^{(\alpha,\beta)} := \{P \in \mathbb{S}_n^{(\alpha,\beta)} : \varepsilon_{\alpha\beta}(P) > \lambda_1\},$$
$$\mathcal{L}_{m,n}^{(\alpha,\beta)} := \{P \in \mathbb{S}_{m,n}^{(\alpha,\beta)} : \varepsilon_{\alpha\beta}(P) > \lambda_1\},$$
$$\mathcal{L}_{\mathcal{R},n}^{(\alpha,\beta)} := \{P \in \mathbb{S}_{\mathcal{R},n}^{(\alpha,\beta)} : \varepsilon_{\alpha\beta}(P) > \lambda_1\},$$

where $\lambda_1 = \frac{\beta-\alpha}{2+\alpha+\beta}$ corresponds to the sole root of the Jacobi polynomial $p_1^{(\alpha,\beta)}$, the following proposition holds.

Proposition 3.7.
If the sets $\mathcal{L}_n$, $\mathcal{L}_n^m$ and $\mathcal{L}_n^{\mathcal{R}}$ are nonempty, then

$$\arg\min_{P\in\mathcal{L}_n^{(\alpha,\beta)}} \mathrm{var}_S^{\alpha\beta}(P) = \arg\max_{P\in\mathcal{L}_n^{(\alpha,\beta)}} \varepsilon_{\alpha\beta}(P) = \mathcal{P}_n^{(\alpha,\beta)},$$

$$\arg\min_{P\in\mathcal{L}_{m,n}^{(\alpha,\beta)}} \mathrm{var}_S^{\alpha\beta}(P) = \arg\max_{P\in\mathcal{L}_{m,n}^{(\alpha,\beta)}} \varepsilon_{\alpha\beta}(P) = \mathcal{P}_{m,n}^{(\alpha,\beta)},$$

$$\arg\min_{P\in\mathcal{L}_{\mathcal{R},n}^{(\alpha,\beta)}} \mathrm{var}_S^{\alpha\beta}(P) = \arg\max_{P\in\mathcal{L}_{\mathcal{R},n}^{(\alpha,\beta)}} \varepsilon_{\alpha\beta}(P) = \mathcal{P}_{\mathcal{R},n}^{(\alpha,\beta)}.$$

Proof. We consider the space variance $\mathrm{var}_S^{\alpha\beta}$ as a function of $\lambda = \varepsilon_{\alpha\beta}(f)$. We have

$$\mathrm{var}_S^{\alpha\beta}(\lambda) = \frac{1-\lambda^2}{(\lambda-\lambda_1)^2},$$

$$\frac{d\,\mathrm{var}_S^{\alpha\beta}}{d\lambda}(\lambda) = \frac{-2(\lambda-\lambda_1)\lambda - 2(1-\lambda^2)}{(\lambda-\lambda_1)^3} = \frac{-2(1-\lambda_1\lambda)}{(\lambda-\lambda_1)^3}.$$

Therefore, the derivative $\frac{d}{d\lambda}\,\mathrm{var}_S^{\alpha\beta}$ is strictly decaying on the open interval $(\lambda_1,1)$ and strictly increasing on $(-1,\lambda_1)$. So, for $P \in \mathcal{L}_n^{(\alpha,\beta)}, \mathcal{L}_{m,n}^{(\alpha,\beta)}, \mathcal{L}_{\mathcal{R},n}^{(\alpha,\beta)}$, maximizing $\varepsilon_{\alpha\beta}(P)$ yields the same result as minimizing $\mathrm{var}_S^{\alpha\beta}(P)$. $\square$

Remark 3.8. Whereas it can not be guaranteed that the sets $\mathcal{L}_{m,n}^{(\alpha,\beta)}$ and $\mathcal{L}_{\mathcal{R},n}^{(\alpha,\beta)}$ are nonempty, the non-emptiness of the sets $\mathcal{L}_n^{(\alpha,\beta)}$, $n \geq 1$, is a consequence of the interlacing property of the zeros of the Jacobi polynomials (cf. [83, Theorem 3.3.2], [10, Theorem 5.3]). Namely, this interlacing property implies that $\varepsilon_{\alpha\beta}(\mathcal{P}_n^{(\alpha,\beta)}) = \lambda_{n+1} > \lambda_n > \ldots > \lambda_1$.

3.1.1. Explicit expression for the optimally space localized polynomials

Our next goal is to find explicit expressions for the optimal polynomials $\mathcal{P}_n^{(\alpha,\beta)}$, $\mathcal{P}_{m,n}^{(\alpha,\beta)}$ and $\mathcal{P}_{\mathcal{R},n}^{(\alpha,\beta)}$ derived in Theorem 3.6. To this end, we need a Christoffel-Darboux type formula for the associated Jacobi polynomials $p_l^{(\alpha,\beta)}(x,m)$ and $p_l^{(\alpha,\beta)}(x,m,\gamma,\delta)$.

Lemma 3.9.
Let $p_l^{(\alpha,\beta)}(x,m)$ and $p_l^{(\alpha,\beta)}(x,m,\gamma,\delta)$ be the associated and the scaled co-recursive associated Jacobi polynomials as defined in (3.12) and (3.13). Then, the following Christoffel-Darboux type formulas hold:

$$\sum_{k=m}^{n} p_k^{(\alpha,\beta)}(x)p_{k-m}^{(\alpha,\beta)}(y,m) \tag{3.22}$$

$$= b_{n+1}\frac{p_{n+1}^{(\alpha,\beta)}(x)p_{n-m}^{(\alpha,\beta)}(y,m) - p_{n-m+1}^{(\alpha,\beta)}(y,m)p_n^{(\alpha,\beta)}(x)}{x-y} + b_m\frac{p_{m-1}^{(\alpha,\beta)}(x)}{x-y},$$

101

$$\sum_{k=m}^{n} p_k^{(\alpha,\beta)}(x) p_{k-m}^{(\alpha,\beta)}(y,m,\gamma,\delta) \tag{3.23}$$

$$= b_{n+1} \frac{p_{n+1}^{(\alpha,\beta)}(x) p_{n-m}^{(\alpha,\beta)}(y,m,\gamma,\delta) - p_{n-m+1}^{(\alpha,\beta)}(y,m,\gamma,\delta) p_n^{(\alpha,\beta)}(x)}{x-y}$$

$$+ \frac{p_m^{(\alpha,\beta)}(x)((\delta-1)y-\gamma)}{x-y} + b_m \frac{p_{m-1}^{(\alpha,\beta)}(x)}{x-y}.$$

Proof. We follow the lines of the proof of the original Christoffel-Darboux formula (see [10, Theorem 4.5]). By (3.9) and (3.12), we have for $k \geq m$ the identities

$$x p_k^{(\alpha,\beta)}(x) p_{k-m}^{(\alpha,\beta)}(y,m)$$
$$= b_{k+1} p_{k+1}^{(\alpha,\beta)}(x) p_{k-m}^{(\alpha,\beta)}(y,m) + a_k p_k^{(\alpha,\beta)}(x) p_{k-m}^{(\alpha,\beta)}(y,m) + b_k p_{k-1}^{(\alpha,\beta)}(x) p_{k-m}^{(\alpha,\beta)}(y,m),$$
$$y p_k^{(\alpha,\beta)}(x) p_{k-m}^{(\alpha,\beta)}(y,m)$$
$$= b_{k+1} p_k^{(\alpha,\beta)}(x) p_{k-m+1}^{(\alpha,\beta)}(y,m) + a_k p_k^{(\alpha,\beta)}(x) p_{k-m}^{(\alpha,\beta)}(y,m) + b_k p_k^{(\alpha,\beta)}(x) p_{k-m-1}^{(\alpha,\beta)}(y,m).$$

Subtracting the second equation from the first, we get

$$(x-y) p_k^{(\alpha,\beta)}(x) p_{k-m}^{(\alpha,\beta)}(y,m)$$
$$= b_{k+1} \big(p_{k+1}^{(\alpha,\beta)}(x) p_{k-m}^{(\alpha,\beta)}(y,m) - p_k^{(\alpha,\beta)}(x) p_{k-m+1}^{(\alpha,\beta)}(y,m) \big)$$
$$- b_k \big(p_k^{(\alpha,\beta)}(x) p_{k-m-1}^{(\alpha,\beta)}(y,m) - p_{k-1}^{(\alpha,\beta)}(x) p_{k-m}^{(\alpha,\beta)}(y,m) \big).$$

Let

$$F_k(x,y) = b_{k+1} \frac{p_{k+1}^{(\alpha,\beta)}(x) p_{k-m}^{(\alpha,\beta)}(y,m) - p_k^{(\alpha,\beta)}(x) p_{k-m+1}^{(\alpha,\beta)}(y,m)}{x-y}.$$

Then, the last equation can be rewritten as

$$p_k^{(\alpha,\beta)}(x) p_{k-m}^{(\alpha,\beta)}(y,m) = F_k(x,y) - F_{k-1}(x,y), \quad k \geq m,$$

where $F_{m-1}(x,y) = -b_m p_{m-1}^{(\alpha,\beta)}(x)$. Summing the latter from m to n, we obtain (3.22). Analogously, we get for the scaled co-recursive associated polynomials

$$p_k^{(\alpha,\beta)}(x) p_{k-m}^{(\alpha,\beta)}(y,m,\gamma,\delta) = G_k(x,y) - G_{k-1}(x,y), \quad k \geq m+1,$$
$$p_m^{(\alpha,\beta)}(x) p_0^{(\alpha,\beta)}(y,m,\gamma,\delta) = p_m^{(\alpha,\beta)}(x),$$

where

$$G_k(x,y) = b_{k+1} \frac{p_{k+1}^{(\alpha,\beta)}(x) p_{k-m}^{(\alpha,\beta)}(y,m,\gamma,\delta) - p_k^{(\alpha,\beta)}(x) p_{k-m+1}^{(\alpha,\beta)}(y,m,\gamma,\delta)}{x-y},$$
$$G_m(x,y) = \frac{b_{m+1} p_{m+1}^{(\alpha,\beta)}(x) - p_m^{(\alpha,\beta)}(x)(\delta y - a_m - \gamma)}{x-y}, \quad k \geq m+1.$$

Then, summing from m to n, we get

$$\sum_{k=m}^{n} p_k^{(\alpha,\beta)}(x)p_{k-m}^{(\alpha,\beta)}(y,m,\gamma,\delta) = \sum_{k=m+1}^{n} (G_k(x,y) - G_{k-1}(x,y)) + p_m^{(\alpha,\beta)}(x)$$

$$=G_n(x,y) - \frac{b_{m+1}p_{m+1}^{(\alpha,\beta)}(x) + p_m^{(\alpha,\beta)}(x)(\delta y - a_m - \gamma)}{x-y} + \frac{p_m^{(\alpha,\beta)}(x)(x-y)}{x-y}$$

$$=G_n(x,y) + \frac{p_m^{(\alpha,\beta)}(x)((\delta-1)y - \gamma)}{x-y} + b_m\frac{p_{m-1}^{(\alpha,\beta)}(x)}{x-y}.$$

Hence, we obtain formula (3.23). $\qquad\qquad\square$

As a direct consequence of the Christoffel-Darboux type formulas in Lemma 3.9, we get the following explicit formulas for the optimal polynomials in Theorem 3.6:

Corollary 3.10.
The optimal polynomials $\mathcal{P}_n^{(\alpha,\beta)}$, $\mathcal{P}_{m,n}^{(\alpha,\beta)}$ and $\mathcal{P}_{\mathcal{R},n}^{(\alpha,\beta)}$ in Theorem 3.6 have the explicit form

$$\mathcal{P}_n^{(\alpha,\beta)}(t) = \kappa_1 b_{n+1} \frac{p_{n+1}^{(\alpha,\beta)}(\cos t)p_n^{(\alpha,\beta)}(\lambda_{n+1})}{\cos t - \lambda_{n+1}},$$

$$\mathcal{P}_{m,n}^{(\alpha,\beta)}(t) = \kappa_2 \frac{b_{n+1}p_{n+1}^{(\alpha,\beta)}(\cos t)p_{n-m}^{(\alpha,\beta)}(\lambda_{n-m+1}^m, m) + b_m p_{m-1}^{(\alpha,\beta)}(\cos t)}{\cos t - \lambda_{n-m+1}^m},$$

$$\mathcal{P}_{\mathcal{R},n}^{(\alpha,\beta)}(t) = \kappa_3 \left(\frac{b_{n+1}p_{n+1}^{(\alpha,\beta)}(\cos t)p_{n-m}^{(\alpha,\beta)}(\lambda_{n-m+1}^{\mathcal{R}}, m, \gamma_{\mathcal{R}}, \delta_{\mathcal{R}})}{\cos t - \lambda_{n-m+1}^{\mathcal{R}}} \right.$$

$$\left. + \frac{p_m^{(\alpha,\beta)}(\cos t)((\delta_{\mathcal{R}}-1)\lambda_{n-m+1}^{\mathcal{R}} - \gamma_{\mathcal{R}}) + b_m p_{m-1}^{(\alpha,\beta)}(\cos t)}{\cos t - \lambda_{n-m+1}^{\mathcal{R}}} \right),$$

where the constants κ_1, κ_2, κ_3 and the roots λ_{n+1} λ_{n-m+1}^m and $\lambda_{n-m+1}^{\mathcal{R}}$ are given as in Theorem 3.6.

Example 3.11. We consider the orthonormal Chebyshev polynomials t_n corresponding to the Jacobi polynomials $p_n^{(\alpha,\beta)}$ with $\alpha = \beta = -\frac{1}{2}$ and the weight function $w_{\alpha\beta}(t) = 1$. The orthonormal Chebyshev polynomials are explicitly given as (see [25, p. 28-29])

$$t_0(\cos t) = \frac{1}{\sqrt{\pi}}, \quad t_n(\cos t) = \sqrt{\frac{2}{\pi}}\cos(nt), \quad n \geq 1.$$

The largest zero of the Chebyshev polynomials t_{n+1} is given by $\lambda_{n+1} = \cos(\frac{\pi}{2n+2})$ (see [83, (6.3.5)]). The normalized associated polynomials $t_n(x,m)$, $m \geq 1$, correspond to the Chebyshev polynomials u_n of the second kind given by (see [25, p. 28-29])

$$u_n(\cos t) = \sqrt{\frac{2}{\pi}}\frac{\sin((n+1)t)}{\sin t}, \quad n \geq 0.$$

The largest zero of the polynomials u_{n+1} is given by $\lambda_{n+1} = \cos(\frac{\pi}{n+2})$. So, in the case of the Chebyshev polynomials, we get for the optimally space localized polynomials $\mathcal{P}_n^{(-\frac{1}{2},-\frac{1}{2})}$ and $\mathcal{P}_{m,n}^{(-\frac{1}{2},-\frac{1}{2})}$, $n \geq 1$, the formulas

$$\mathcal{P}_n^{(-\frac{1}{2},-\frac{1}{2})}(t) = \frac{\kappa_1}{\pi}\left(1 + 2\sum_{k=1}^{n}\cos\left(\frac{k\pi}{2n+2}\right)\cos(kt)\right) = \frac{\kappa_1}{\pi}\frac{\cos((n+1)t)\cos(\frac{n\pi}{2n+2})}{\cos t - \cos(\frac{\pi}{2n+2})}.$$

$$\mathcal{P}_{m,n}^{(-\frac{1}{2},-\frac{1}{2})}(t) = \frac{2\kappa_2}{\pi\sin(\frac{\pi}{n-m+2})}\left(\sum_{k=m}^{n}\sin\left(\frac{(k-m+1)\pi}{n-m+2}\right)\cos(kt)\right), \quad m \geq 1.$$

The polynomials $\mathcal{P}_n^{(-\frac{1}{2},-\frac{1}{2})}$ are almost identical to the Rogosinski kernel R_n which is defined as

$$R_n(t) = 1 + 2\sum_{k=1}^{n}\cos\left(\frac{k\pi}{2n+1}\right)\cos(kt), \quad t \in [0,\pi].$$

For more details on the Rogosinski kernel and the relation to the optimal polynomials $\mathcal{P}_n^{(-\frac{1}{2},-\frac{1}{2})}$, we refer to [50, p. 112-114], [70, Section 5.2] and [71].

$\mathcal{P}_6^{(-\frac{1}{2},-\frac{1}{2})}(t).$ $\mathcal{P}_{12}^{(-\frac{1}{2},-\frac{1}{2})}(t).$ $\mathcal{P}_{24}^{(-\frac{1}{2},-\frac{1}{2})}(t).$

$\mathcal{P}_{6,12}^{(-\frac{1}{2},-\frac{1}{2})}(t).$ $\mathcal{P}_{6,18}^{(-\frac{1}{2},-\frac{1}{2})}(t).$ $\mathcal{P}_{6,30}^{(-\frac{1}{2},-\frac{1}{2})}(t).$

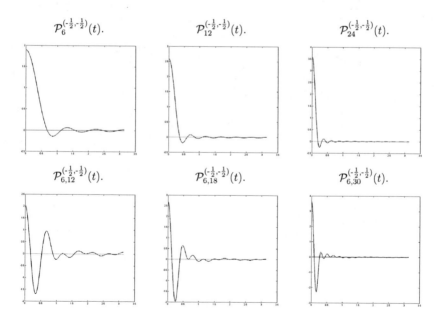

Figure 13: Optimally space localized polynomials and wavelets for Chebyshev expansions on $[0,\pi]$ ($\alpha = \beta = -\frac{1}{2}$)

$\mathcal{P}_6^{(0.42,0.42)}(t).$ $\mathcal{P}_{12}^{(0.42,0.42)}(t).$ $\mathcal{P}_{24}^{(0.42,0.42)}(t).$

$\mathcal{P}_{6,12}^{(0.42,0.42)}(t).$ $\mathcal{P}_{6,18}^{(0.42,0.42)}(t).$ $\mathcal{P}_{6,30}^{(0.42,0.42)}(t).$

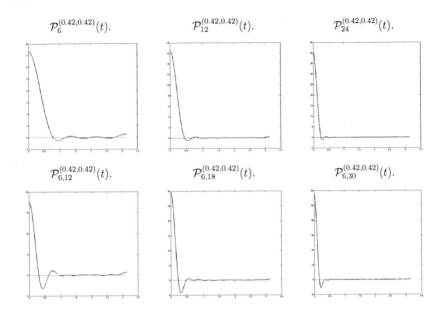

Figure 14: Optimally space localized polynomials and wavelets for Jacobi expansions on $[0, \pi]$ with $\alpha = \beta = 0.42$

3.1.2. Space-frequency localization of the optimally space localized polynomials

In this section, we will compute the frequency variance $\mathrm{var}_F^{\alpha\beta}(\mathcal{P}_n^{(\alpha,\beta)})$ of the optimally space localized polynomials $\mathcal{P}_n^{(\alpha,\beta)}$. This will enable us to determine the space-frequency localization of the polynomials $\mathcal{P}_n^{(\alpha,\beta)}$ and, in particular, to determine the asymptotic behavior of the uncertainty product $\mathrm{var}_S^{\alpha\beta}(\mathcal{P}_n^{(\alpha,\beta)}) \cdot \mathrm{var}_F^{\alpha\beta}(\mathcal{P}_n^{(\alpha,\beta)})$ as the degree n of the polynomial $\mathcal{P}_n^{(\alpha,\beta)}$ tends to infinity. Again, we need a Christoffel-Darboux type formula, but this time for the derivatives $p_k^{(\alpha,\beta)'}(x) := \frac{d}{dx}p_k^{(\alpha,\beta)}(x)$, $p_k^{(\alpha,\beta)''}(x) := \frac{d^2}{dx^2}p_k^{(\alpha,\beta)}(x)$ of the Jacobi polynomial $p_k^{(\alpha,\beta)}$.

Lemma 3.12.
The following Christoffel-Darboux type formulas hold:

$$\sum_{k=1}^{n} p_k^{(\alpha,\beta)'}(x)p_k^{(\alpha,\beta)'}(y) + \sum_{k=1}^{n} \frac{p_k^{(\alpha,\beta)}(x)p_k^{(\alpha,\beta)'}(y) - p_k^{(\alpha,\beta)'}(x)p_k^{(\alpha,\beta)}(y)}{x - y}$$
$$= b_{n+1} \frac{p_{n+1}^{(\alpha,\beta)'}(x)p_n^{(\alpha,\beta)'}(y) - p_{n+1}^{(\alpha,\beta)'}(y)p_n^{(\alpha,\beta)'}(x)}{x - y}, \tag{3.24}$$

105

$$\sum_{l=1}^{n} p_l^{(\alpha,\beta)'}(x)p_l^{(\alpha,\beta)}(x) = \frac{1}{2}b_{n+1}\left(p_{n+1}^{(\alpha,\beta)''}(x)p_n^{(\alpha,\beta)}(x) - p_n^{(\alpha,\beta)''}(x)p_{n+1}^{(\alpha,\beta)}(x)\right), \tag{3.25}$$

$$\sum_{k=1}^{n} p_k^{(\alpha,\beta)''}(x)p_k^{(\alpha,\beta)}(x) = \frac{1}{3}b_{n+1}\left(p_{n+1}^{(\alpha,\beta)'''}(x)p_n^{(\alpha,\beta)}(x) - p_{n+1}^{(\alpha,\beta)}(x)p_n^{(\alpha,\beta)'''}(x)\right), \tag{3.26}$$

$$\sum_{k=1}^{n} p_k^{(\alpha,\beta)'}(x)^2 = \frac{1}{6}b_{n+1}\left(p_{n+1}^{(\alpha,\beta)}(x)p_n^{(\alpha,\beta)'''}(x) - 3p_{n+1}^{(\alpha,\beta)'}(x)p_n^{(\alpha,\beta)''}(x)\right.$$
$$\left. + 3p_{n+1}^{(\alpha,\beta)''}(x)p_n^{(\alpha,\beta)'}(x) - p_{n+1}^{(\alpha,\beta)'''}(x)p_n^{(\alpha,\beta)}(x)\right). \tag{3.27}$$

Proof. In principle, we follow again the lines of the proof of Theorem 4.5 in [10]. By the three-term recurrence relation of the orthonormal Jacobi polynomials (3.9), we have the identities

$$xp_k^{(\alpha,\beta)'}(x)p_k^{(\alpha,\beta)'}(y) = \left((xp_k^{(\alpha,\beta)})'(x) - p_k^{(\alpha,\beta)}(x)\right)p_k^{(\alpha,\beta)'}(y)$$
$$= b_{k+1}p_{k+1}^{(\alpha,\beta)'}(x)p_k^{(\alpha,\beta)'}(y) + a_k p_k^{(\alpha,\beta)'}(x)p_k^{(\alpha,\beta)'}(y)$$
$$+ b_k p_{k-1}^{(\alpha,\beta)'}(x)p_k^{(\alpha,\beta)'}(y) - p_k^{(\alpha,\beta)}(x)p_k^{(\alpha,\beta)'}(y),$$
$$yp_k^{(\alpha,\beta)'}(x)p_k^{(\alpha,\beta)'}(y) = p_k^{(\alpha,\beta)'}(x)\left((yp_k^{(\alpha,\beta)})'(y) - p_k^{(\alpha,\beta)}(y)\right)$$
$$= b_{k+1}p_k^{(\alpha,\beta)'}(x)p_{k+1}^{(\alpha,\beta)'}(y) + a_k p_k^{(\alpha,\beta)'}(x)p_k^{(\alpha,\beta)'}(y)$$
$$+ b_k p_k^{(\alpha,\beta)'}(x)p_{k-1}^{(\alpha,\beta)'}(y) - p_k^{(\alpha,\beta)'}(x)p_k^{(\alpha,\beta)}(y).$$

Subtracting the second equation from the first, we get

$$(x-y)p_k^{(\alpha,\beta)'}(x)p_k^{(\alpha,\beta)'}(y) = b_{k+1}\left(p_{k+1}^{(\alpha,\beta)'}(x)p_k^{(\alpha,\beta)'}(y) - p_k^{(\alpha,\beta)'}(x)p_{k+1}^{(\alpha,\beta)'}(y)\right)$$
$$- b_k\left(p_k^{(\alpha,\beta)'}(x)p_{k-1}^{(\alpha,\beta)'}(y) - p_{k-1}^{(\alpha,\beta)'}(x)p_k^{(\alpha,\beta)'}(y)\right)$$
$$- \left(p_k^{(\alpha,\beta)}(x)p_k^{(\alpha,\beta)'}(y) - p_k^{(\alpha,\beta)'}(x)p_k^{(\alpha,\beta)}(y)\right).$$

Now, summing from $k = 1$ to $k = n$ and dividing by $(x - y)$, we get equation (3.24).

For the limit $x \to y$ in (3.24), we get the equation

$$2\sum_{k=1}^{n} p_k^{(\alpha,\beta)'}(x)^2 - \sum_{k=1}^{n} p_k^{(\alpha,\beta)''}(x)p_k^{(\alpha,\beta)}(x)$$
$$= b_{n+1}\left(p_{n+1}^{(\alpha,\beta)''}(x)p_n^{(\alpha,\beta)'}(x) - p_{n+1}^{(\alpha,\beta)'}(x)p_n^{(\alpha,\beta)''}(x)\right). \tag{3.28}$$

Further, we have the well known Christoffel-Darboux formula [10, Theorem 4.6]

$$\sum_{l=0}^{n} p_l^{(\alpha,\beta)}(x)^2 = b_{n+1}\left(p_{n+1}^{(\alpha,\beta)'}(x)p_n^{(\alpha,\beta)}(x) - p_n^{(\alpha,\beta)'}(x)p_{n+1}^{(\alpha,\beta)}(x)\right). \tag{3.29}$$

Differentiating both sides of (3.29) twice with respect to x, we get

$$2\sum_{l=1}^{n} p_l^{(\alpha,\beta)'}(x)p_l^{(\alpha,\beta)}(x) = b_{n+1}\left(p_{n+1}^{(\alpha,\beta)''}(x)p_n^{(\alpha,\beta)}(x) - p_n^{(\alpha,\beta)''}(x)p_{n+1}^{(\alpha,\beta)}(x)\right)$$

$$2\sum_{l=1}^{n}\left(p_l^{(\alpha,\beta)''}(x)p_l^{(\alpha,\beta)}(x) + p_l^{(\alpha,\beta)'}(x)^2\right) = b_{n+1}\left(p_{n+1}^{(\alpha,\beta)'''}(x)p_n^{(\alpha,\beta)}(x) - p_n^{(\alpha,\beta)'''}(x)p_{n+1}^{(\alpha,\beta)}(x)\right.$$
$$\left. + p_{n+1}^{(\alpha,\beta)''}(x)p_n^{(\alpha,\beta)'}(x) - p_n^{(\alpha,\beta)''}(x)p_{n+1}^{(\alpha,\beta)'}(x)\right).$$

Then, the first equation above gives (3.25) and the second equation in combination with formula (3.28) implies the equations (3.26) and (3.27). $\square$

With the help of the Christoffel-Darboux type formulas in Lemma 3.12, it is possible to compute the frequency variance of the optimally space localized polynomials $\mathcal{P}_n^{(\alpha,\beta)}$ explicitly. To this end, we need also the second-order differential operator $L_{\alpha\beta}$ and the respective differential equation of the Jacobi polynomials (see equations (1.66) and (1.67)), i.e.,

$$L_{\alpha\beta}p_n^{(\alpha,\beta)} = -n(n+\alpha+\beta+1)p_n^{(\alpha,\beta)}, \qquad (3.30)$$

where the differential operator $L_{\alpha\beta}$ is given in the variable $x = \cos t$ as

$$L_{\alpha\beta} = (1-x^2)\frac{d^2}{dx^2} + (\beta - \alpha - (\alpha+\beta+2)x)\frac{d}{dx}. \qquad (3.31)$$

Proposition 3.13.
The frequency variance $\mathrm{var}_F^{\alpha\beta}$ *of the optimal polynomial* $\mathcal{P}_n^{(\alpha,\beta)}$ *has the explicit form*

$$\mathrm{var}_F^{\alpha\beta}(\mathcal{P}_n^{(\alpha,\beta)}) = \frac{n(n+\alpha+\beta+3)}{3} + \frac{(\alpha-\beta+\lambda_{n+1}(\alpha+\beta))^2 - 4(\lambda_{n+1})^2}{6(1-(\lambda_{n+1})^2)}. \qquad (3.32)$$

Proof. By Corollary 3.10 and Theorem 3.6, the optimal polynomial $\mathcal{P}_n^{(\alpha,\beta)}$ has the representations

$$\mathcal{P}_n^{(\alpha,\beta)}(t) = \kappa_1 b_{n+1}\frac{p_{n+1}^{(\alpha,\beta)}(\cos t)p_n^{(\alpha,\beta)}(\lambda_{n+1})}{\cos t - \lambda_{n+1}} = \kappa_1\sum_{l=0}^{n}p_l^{(\alpha,\beta)}(\lambda_{n+1})p_l^{(\alpha,\beta)}(\cos t).$$

Without loss of generality we can assume that $\mathcal{P}_n^{(\alpha,\beta)}(0) > 0$. Then, the constant κ_1 is given by

$$\kappa_1 = \|\mathcal{P}_n^{(\alpha,\beta)}\|_{w_{\alpha\beta}}^{-1} = \left(\sum_{l=0}^{n}p_l^{(\alpha,\beta)}(\lambda_{n+1})^2\right)^{-\frac{1}{2}}.$$

By formula (1.70) and the equations (3.31) and (3.30), we get for the frequency variance of $\mathcal{P}_n^{(\alpha,\beta)}$:

$$
\mathrm{var}_F^{\alpha\beta}(\mathcal{P}_n^{(\alpha,\beta)}) \overset{(1.70)}{=} \langle -L_{\alpha,\beta}\mathcal{P}_n^{(\alpha,\beta)}, \mathcal{P}_n^{(\alpha,\beta)} \rangle_{w_{\alpha\beta}}
$$

$$
\overset{(3.30)}{=} \kappa_1^2 \sum_{l=0}^{n} l(l+\alpha+\beta+1)p_l^{(\alpha,\beta)}(\lambda_{n+1})^2
$$

$$
\overset{(3.30)}{=} -\kappa_1^2 \sum_{l=0}^{n} (L_{\alpha\beta}p_l^{(\alpha,\beta)})(\lambda_{n+1})p_l^{(\alpha,\beta)}(\lambda_{n+1})
$$

$$
\overset{(3.31)}{=} -(1-(\lambda_{n+1})^2)\frac{\sum_{l=0}^{n} p_l^{(\alpha,\beta)''}(\lambda_{n+1})p_l^{(\alpha,\beta)}(\lambda_{n+1})}{\sum_{l=0}^{n} p_l^{(\alpha,\beta)}(\lambda_{n+1})^2}
$$

$$
+ (\alpha-\beta+\lambda_{n+1}(\alpha+\beta+2))\frac{\sum_{l=0}^{n} p_l^{(\alpha,\beta)'}(\lambda_{n+1})p_l^{(\alpha,\beta)}(\lambda_{n+1})}{\sum_{l=0}^{n} p_l^{(\alpha,\beta)}(\lambda_{n+1})^2}.
$$

Now, using Lemma 3.12 and the fact that λ_{n+1} is the largest zero of $p_{n+1}^{(\alpha,\beta)}$, we get

$$
\mathrm{var}_F^{\alpha\beta}(\mathcal{P}_n^{(\alpha,\beta)}) = -\frac{1-(\lambda_{n+1})^2}{3}\frac{p_{n+1}^{(\alpha,\beta)'''}(\lambda_{n+1})}{p_{n+1}^{(\alpha,\beta)'}(\lambda_{n+1})} + \frac{\alpha-\beta+\lambda_{n+1}(\alpha+\beta+2)}{2}\frac{p_{n+1}^{(\alpha,\beta)''}(\lambda_{n+1})}{p_{n+1}^{(\alpha,\beta)'}(\lambda_{n+1})}.
$$

The derivative $P_{n+1}^{(\alpha,\beta)'}$ is related to the Jacobi polynomial $P_n^{(\alpha+1,\beta+1)}$ by (cf. [83, (4.21.7)])

$$
P_{n+1}^{(\alpha,\beta)'}(x) = \frac{n+\alpha+\beta+1}{2}P_n^{(\alpha+1,\beta+1)}(x). \tag{3.33}
$$

Hence, using (3.33) and the formula (3.31) for the operator $L_{\alpha\beta}$, we get for the frequency variance

$$
\mathrm{var}_F^{\alpha\beta}(\mathcal{P}_n^{(\alpha,\beta)}) = \frac{(\lambda_{n+1})^2-1}{3}\frac{p_n^{(\alpha+1,\beta+1)''}(\lambda_{n+1})}{p_n^{(\alpha+1,\beta+1)}(\lambda_{n+1})} + \frac{\alpha-\beta+\lambda_{n+1}(\alpha+\beta+2)}{2}\frac{p_n^{(\alpha+1,\beta+1)'}(\lambda_{n+1})}{p_n^{(\alpha+1,\beta+1)}(\lambda_{n+1})}
$$

$$
= \frac{n(n+\alpha+\beta+3)}{3} + \frac{\alpha-\beta+\lambda_{n+1}(\alpha+\beta-2)}{6}\frac{p_n^{(\alpha+1,\beta+1)'}(\lambda_{n+1})}{p_n^{(\alpha+1,\beta+1)}(\lambda_{n+1})}
$$

$$
= \frac{n(n+\alpha+\beta+3)}{3} + \frac{\alpha-\beta+\lambda_{n+1}(\alpha+\beta-2)}{6(1-(\lambda_{n+1})^2)}\frac{(1-(\lambda_{n+1})^2)p_{n+1}^{(\alpha,\beta)''}(\lambda_{n+1})}{p_{n+1}^{(\alpha,\beta)'}(\lambda_{n+1})}
$$

$$
= \frac{n(n+\alpha+\beta+3)}{3} + \frac{(\alpha-\beta+\lambda_{n+1}(\alpha+\beta))^2-4(\lambda_{n+1})^2}{6(1-(\lambda_{n+1})^2)}.
$$

$\square$

Finally, we can show that the uncertainty product $\mathrm{var}_S^{\alpha\beta}(\mathcal{P}_n^{(\alpha,\beta)}) \cdot \mathrm{var}_F^{\alpha\beta}(\mathcal{P}_n^{(\alpha,\beta)})$ for the optimal polynomials $\mathcal{P}_n^{(\alpha,\beta)}$ is uniformly bounded for all $n \in \mathbb{N}$. In the case that $|\alpha| = |\beta| = \frac{1}{2}$, we even get explicit results.

Theorem 3.14.

There exists a constant $C_{\alpha\beta}$, independent of n, such that the uncertainty product of the optimal polynomial $\mathcal{P}_n^{(\alpha,\beta)}$ is bounded by

$$\mathrm{var}_S^{\alpha\beta}(\mathcal{P}_n^{(\alpha,\beta)}) \cdot \mathrm{var}_F^{\alpha\beta}(\mathcal{P}_n^{(\alpha,\beta)}) \le C_{\alpha\beta}.$$

Further, if $-\frac{1}{2} < \alpha, \beta < \frac{1}{2}$, then

$$\lim_{n\to\infty} \mathrm{var}_S^{\alpha\beta}(\mathcal{P}_n^{(\alpha,\beta)}) \cdot \mathrm{var}_F^{\alpha\beta}(\mathcal{P}_n^{(\alpha,\beta)}) \le \frac{(\alpha+\beta+2)^2(\pi^2 + 2\alpha^2 - 2)}{12(\alpha+1)^2}.$$

In the case that $|\alpha| = |\beta| = \frac{1}{2}$, we get

$$\lim_{n\to\infty} \mathrm{var}_S^{-\frac{1}{2},-\frac{1}{2}}(\mathcal{P}_n^{(-\frac{1}{2},-\frac{1}{2})}) \cdot \mathrm{var}_F^{-\frac{1}{2},-\frac{1}{2}}(\mathcal{P}_n^{(-\frac{1}{2},-\frac{1}{2})}) = \frac{\pi^2}{12} - \frac{1}{2} \approx 0.3225 > \frac{1}{4},$$

$$\lim_{n\to\infty} \mathrm{var}_S^{\frac{1}{2},\frac{1}{2}}(\mathcal{P}_n^{(\frac{1}{2},\frac{1}{2})}) \cdot \mathrm{var}_F^{\frac{1}{2},\frac{1}{2}}(\mathcal{P}_n^{(\frac{1}{2},\frac{1}{2})}) = \frac{\pi^2}{3} - \frac{1}{2} \approx 2.7899 > \frac{9}{4},$$

$$\lim_{n\to\infty} \mathrm{var}_S^{\frac{1}{2},-\frac{1}{2}}(\mathcal{P}_n^{(\frac{1}{2},-\frac{1}{2})}) \cdot \mathrm{var}_F^{\frac{1}{2},-\frac{1}{2}}(\mathcal{P}_n^{(\frac{1}{2},-\frac{1}{2})}) = \frac{4}{9}\left(\frac{\pi^2}{3} - \frac{1}{2}\right) \approx 1.2399 > 1,$$

$$\lim_{n\to\infty} \mathrm{var}_S^{-\frac{1}{2},\frac{1}{2}}(\mathcal{P}_n^{(-\frac{1}{2},\frac{1}{2})}) \cdot \mathrm{var}_F^{-\frac{1}{2},\frac{1}{2}}(\mathcal{P}_n^{(-\frac{1}{2},\frac{1}{2})}) = \frac{\pi^2}{3} - 2 \approx 1,2899 > 1.$$

Proof. By [83, Theorem 8.9.1], there exists a constant $c_{\alpha\beta} > -\pi$, independent of n, such that

$$\lambda_{n+1} \ge \cos\left(\frac{\pi + c_{\alpha\beta}}{n+1}\right).$$

Then, we get by Proposition 3.13

$$\mathrm{var}_S^{\alpha\beta}(\mathcal{P}_n^{(\alpha,\beta)}) \cdot \mathrm{var}_F^{\alpha\beta}(\mathcal{P}_n^{(\alpha,\beta)})$$

$$= \frac{1 - (\lambda_{n+1})^2}{\left(\frac{\alpha-\beta}{\alpha+\beta+2} + \lambda_{n+1}\right)^2}\left(\frac{n(n+\alpha+\beta+3)}{3} + \frac{(\alpha-\beta+\lambda_{n+1}(\alpha+\beta))^2 - 4(\lambda_{n+1})^2}{6(1-(\lambda_{n+1})^2)}\right)$$

$$\le \frac{(\pi + c_{\alpha\beta})^2}{\left(\frac{\alpha-\beta}{\alpha+\beta+2} + \lambda_{n+1}\right)^2}\frac{n(n+\alpha+\beta+3)}{3(n+1)^2} + \frac{(\alpha+\beta+2)^2(\alpha-\beta+\lambda_{n+1}(\alpha+\beta-2))}{6(\alpha-\beta+\lambda_{n+1}(\alpha+\beta+2))}.$$

Both terms on the right hand side of the above inequality can be bounded uniformly by a constant independent of n. Hence also the product $\mathrm{var}_S^{\alpha\beta}(\mathcal{P}_n^{(\alpha,\beta)}) \cdot \mathrm{var}_F^{\alpha\beta}(\mathcal{P}_n^{(\alpha,\beta)})$ is uniformly bounded by a constant $C_{\alpha\beta}$.

If $-\frac{1}{2} < \alpha, \beta < \frac{1}{2}$, then by [83, Theorem 6.3.2], the largest zero of $p_{n+1}^{(\alpha,\beta)}(x)$ is bounded by

$$\lambda_{n+1} > \cos\left(\frac{2\pi}{2n+\alpha+\beta+3}\right).$$

Hence,

$$\lim_{n\to\infty} \text{var}_S^{\alpha\beta}(\mathcal{P}_n^{(\alpha,\beta)}) \cdot \text{var}_F^{\alpha\beta}(\mathcal{P}_n^{(\alpha,\beta)})$$

$$= \lim_{n\to\infty} \frac{1-(\lambda_{n+1})^2}{\left(\frac{\alpha-\beta}{\alpha+\beta+2}+\lambda_{n+1}\right)^2}\left(\frac{n(n+\alpha+\beta+3)}{3} + \frac{(\alpha-\beta+\lambda_{n+1}(\alpha+\beta))^2 - 4(\lambda_{n+1})^2}{6(1-(\lambda_{n+1})^2)}\right)$$

$$\leq \lim_{n\to\infty} \frac{4\pi^2}{\left(\frac{\alpha-\beta}{\alpha+\beta+2}+\lambda_{n+1}\right)^2} \frac{n(n+\alpha+\beta+3)}{3(2n+\alpha+\beta+3)^2} + \frac{(\alpha+\beta+2)^2(\alpha-\beta+\lambda_{n+1}(\alpha+\beta-2))}{6(\alpha-\beta+\lambda_{n+1}(\alpha+\beta+2))}$$

$$= \frac{(\alpha+\beta+2)^2(\pi^2+2\alpha^2-2)}{12(\alpha+1)^2}.$$

Finally, if $\alpha=\beta=-\frac{1}{2}$, $\alpha=\beta=\frac{1}{2}$, $\alpha=-\beta=-\frac{1}{2}$, $\alpha=-\beta=\frac{1}{2}$, the extremal zeros of the Jacobi polynomials $p_{n+1}^{(\alpha,\beta)}(x)$ can be computed as (see [83, (6.3.5)])

$$\lambda_{n+1} = \cos\left(\frac{\pi}{2n+2}\right), \quad \lambda_{n+1} = \cos\left(\frac{\pi}{n+1}\right),$$

$$\lambda_{n+1} = \cos\left(\frac{2\pi}{2n+1}\right), \quad \lambda_{n+1} = \cos\left(\frac{\pi}{2n+1}\right),$$

respectively. Therefore, we get

$$\lim_{n\to\infty} \text{var}_S^{\frac{1}{2},\frac{1}{2}}(\mathcal{P}_n^{(\frac{1}{2},\frac{1}{2})}) \cdot \text{var}_F^{\frac{1}{2},\frac{1}{2}}(\mathcal{P}_n^{(\frac{1}{2},\frac{1}{2})}) = \lim_{n\to\infty} \frac{1-(\lambda_{n+1})^2}{(\lambda_{n+1})^2}\frac{n(n+4)}{3} - \frac{1}{2} = \frac{\pi^2}{3} - \frac{1}{2},$$

$$\lim_{n\to\infty} \text{var}_S^{-\frac{1}{2},-\frac{1}{2}}(\mathcal{P}_n^{(-\frac{1}{2},-\frac{1}{2})}) \cdot \text{var}_F^{-\frac{1}{2},-\frac{1}{2}}(\mathcal{P}_n^{(-\frac{1}{2},-\frac{1}{2})}) = \lim_{n\to\infty} \frac{1-(\lambda_{n+1})^2}{(\lambda_{n+1})^2}\frac{n(n+2)}{3} - \frac{1}{2} = \frac{\pi^2}{12} - \frac{1}{2},$$

$$\lim_{n\to\infty} \text{var}_S^{-\frac{1}{2},\frac{1}{2}}(\mathcal{P}_n^{(-\frac{1}{2},\frac{1}{2})}) \cdot \text{var}_F^{-\frac{1}{2},\frac{1}{2}}(\mathcal{P}_n^{(-\frac{1}{2},\frac{1}{2})}) = \lim_{n\to\infty} \frac{1-(\lambda_{n+1})^2}{(\lambda_{n+1}-\frac{1}{2})^2}\frac{n(n+3)}{3} - \frac{2\lambda_{n+1}+1}{3(4\lambda_{n+1}-2)}$$

$$= \frac{\pi^2}{3} - 2,$$

$$\lim_{n\to\infty} \text{var}_S^{\frac{1}{2},-\frac{1}{2}}(\mathcal{P}_n^{(\frac{1}{2},-\frac{1}{2})}) \cdot \text{var}_F^{\frac{1}{2},-\frac{1}{2}}(\mathcal{P}_n^{(\frac{1}{2},-\frac{1}{2})}) = \lim_{n\to\infty} \frac{1-(\lambda_{n+1})^2}{(\frac{1}{2}+\lambda_{n+1})^2}\frac{n(n+3)}{3} + \frac{2(1-2\lambda_{n+1})}{3(1+2\lambda_{n+1})}$$

$$= \frac{4}{9}\left(\frac{\pi^2}{3} - \frac{1}{2}\right).$$

$\square$

3.1.3. Space-frequency localization of the Christoffel-Darboux kernel

We are now going to compare the space-frequency localization of the space optimal polynomials $\mathcal{P}_n^{(\alpha,\beta)}$ with the space-frequency behavior of other well-known families of polynomials. As a first example, we consider the Christoffel-Darboux kernels $K_n^{(\alpha,\beta)}$ of degree n which are defined on $[0,\pi]$ as

$$K_n^{(\alpha,\beta)}(t) := \sum_{l=0}^{n} p_l^{(\alpha,\beta)}(1)p_l^{(\alpha,\beta)}(\cos t), \tag{3.34}$$

and have the explicit form (cf. [83, (4.5.3)])

$$K_n^{(\alpha,\beta)}(t) = 2^{-\alpha-\beta-2}\frac{\Gamma(n+\alpha+\beta+2)}{\Gamma(\alpha+1)\Gamma(n+\beta+1)}P_n^{(\alpha+1,\beta)}(\cos t). \tag{3.35}$$

In the case that $\alpha = \beta = -\frac{1}{2}$, i.e., for the Chebyshev polynomials of first kind, the Christoffel-Darboux kernel corresponds to the Dirichlet kernel D_n given by

$$D_n(t) := K_n^{(-\frac{1}{2},-\frac{1}{2})}(t) = \frac{1}{\pi}\frac{\sin(\frac{2n+1}{2}t)}{\sin(\frac{t}{2})}.$$

For the Dirichlet kernel it is known that the uncertainty product is far from being optimal. More precisely, in [68] it was shown that for the normalized Dirichlet kernel $\tilde{D}_n := D_n/\|D_n\|$ the following formula holds:

$$\mathrm{var}_S(\tilde{D}_n) \cdot \mathrm{var}_F(\tilde{D}_n) = \frac{(4n+1)(n+1)}{12n}.$$

So, the uncertainty product tends linearly to infinity as the degree n of the Dirichlet kernel $\tilde{D}_n$ tends to infinity. A similar result can be shown also for the normalized Christoffel-Darboux kernel $\tilde{K}_n^{(\alpha,\beta)}$ defined by

$$\tilde{K}_n^{(\alpha,\beta)}(\cos t) := \frac{K_n^{(\alpha,\beta)}(\cos t)}{\|K_n^{(\alpha,\beta)}\|_{w_{\alpha\beta}}} = \frac{P_n^{(\alpha+1,\beta)}(\cos t)}{\|P_n^{(\alpha+1,\beta)}\|_{w_{\alpha\beta}}}.$$

Theorem 3.15.
For the normalized Christoffel-Darboux kernel $\tilde{K}_n^{(\alpha,\beta)}$, the following formulas hold:

$$\varepsilon_{\alpha\beta}\left(\tilde{K}_n^{(\alpha,\beta)}\right) = 1 - \frac{2(\alpha+1)}{2n+\alpha+\beta+2},$$

$$\mathrm{var}_F^{\alpha\beta}\left(\tilde{K}_n^{(\alpha,\beta)}\right) = \frac{\alpha+1}{\alpha+2}n(n+\alpha+\beta+2),$$

$$\mathrm{var}_S^{\alpha\beta}\left(\tilde{K}_n^{(\alpha,\beta)}\right) \cdot \mathrm{var}_F^{\alpha\beta}\left(\tilde{K}_n^{(\alpha,\beta)}\right) = \frac{(\alpha+\beta+2)^2}{\alpha+2}\frac{(2n+\beta+1)(n+\alpha+\beta+2)}{4n}.$$

Proof. For the norm $\|P_n^{(\alpha+1,\beta)}\|_{w_{\alpha+1,\beta}}$, we know from formula (1.64) that

$$\|P_n^{(\alpha+1,\beta)}\|_{w_{\alpha+1,\beta}}^2 = \int_0^\pi P_n^{(\alpha+1,\beta)}(\cos t)^2 w_{\alpha+1,\beta}(t)dt$$

$$= \frac{2^{\alpha+\beta+2}\Gamma(\alpha+n+2)\Gamma(\beta+n+1)}{n!\Gamma(\alpha+\beta+n+2)(\alpha+\beta+2n+2)}.$$

To compute the norm $\|P_n^{(\alpha+1,\beta)}\|_{w_{\alpha,\beta}}$, we use first of all the coordinate transform $x = \cos t$. Then, $\|P_n^{(\alpha+1,\beta)}\|_{w_{\alpha,\beta}}^2$ reads as

$$\|P_n^{(\alpha+1,\beta)}\|_{w_{\alpha,\beta}}^2 = \int_{-1}^1 P_n^{(\alpha+1,\beta)}(x)^2(1-x)^\alpha(1+x)^\beta dx.$$

Now, using the definition (1.63) of the polynomial $P_n^{(\alpha+1,\beta)}$, i.e.,

$$P_n^{(\alpha+1,\beta)}(x) = \frac{\Gamma(n+\alpha+2)}{n!\Gamma(n+\alpha+\beta+2)} \sum_{j=0}^n \binom{n}{j} \frac{\Gamma(n+j+\alpha+\beta+2)}{\Gamma(j+\alpha+2)} \left(\frac{x-1}{2}\right)^j, \quad (3.36)$$

and the orthogonality relation (1.64) of the Jacobi polynomials, we can derive

$$\|P_n^{(\alpha+1,\beta)}\|_{w_{\alpha,\beta}}^2 = \frac{\Gamma(n+\alpha+2)}{n!\Gamma(\alpha+2)} \int_{-1}^1 P_n^{(\alpha+1,\beta)}(x)(1-x)^\alpha(1+x)^\beta dx.$$

Next, applying the Rodriguez formula (see [83, 4.3.1]) of the Jacobi polynomial $P_n^{(\alpha+1,\beta)}(x)$ and integrating by parts n times, yields the equation

$$\|P_n^{(\alpha+1,\beta)}\|_{w_{\alpha,\beta}}^2 = \frac{(-1)^n \Gamma(n+\alpha+2)}{(n!)^2 2^n \Gamma(\alpha+2)} \int_{-1}^1 \left(\frac{d}{dx}\right)^{(n)} \left[(1-x)^{n+\alpha+1}(1+x)^{\beta+n}\right] \frac{1}{1-x} dx$$

$$= \frac{\Gamma(n+\alpha+2)}{n! 2^n \Gamma(\alpha+2)} \int_{-1}^1 (1-x)^\alpha(1+x)^{\beta+n} dx$$

$$= \frac{\Gamma(n+\alpha+2)}{n!\Gamma(\alpha+2)} 2^{\alpha+\beta+1} \frac{\Gamma(\alpha+1)\Gamma(\beta+n+1)}{\Gamma(\alpha+\beta+n+2)}$$

$$= \frac{2^{\alpha+\beta+1}}{\alpha+1} \frac{\Gamma(n+\alpha+2)\Gamma(\beta+n+1)}{\Gamma(\alpha+\beta+n+2)n!}, \quad (3.37)$$

where in the penultimate equality we used the integral formula (1.61). In total, we get for the mean value $\varepsilon_{\alpha\beta}$ of the normalized Christoffel-Darboux kernel:

$$1 - \varepsilon_{\alpha\beta}\left(\tilde{K}_n^{(\alpha,\beta)}\right) = 1 - \frac{\varepsilon_{\alpha\beta}(P_n^{(\alpha+1,\beta)})}{\|P_n^{(\alpha+1,\beta)}\|_{w_{\alpha\beta}}^2} = \frac{\int_0^\pi (1-\cos t)P_n^{(\alpha+1,\beta)}(\cos t)^2 w_{\alpha,\beta}(t)dt}{\|P_n^{(\alpha+1,\beta)}\|_{w_{\alpha\beta}}^2}$$

$$= \frac{\|P_n^{(\alpha+1,\beta)}\|_{w_{\alpha+1,\beta}}^2}{\|P_n^{(\alpha+1,\beta)}\|_{w_{\alpha\beta}}^2} = \frac{2(\alpha+1)}{2n+\alpha+\beta+2}.$$

Next, using the representation (3.36) for the polynomial $P_{n-1}^{(\alpha+2,\beta+1)}$ and the orthogonality relation (1.64), we can deduce the following identity

$$\int_{-1}^1 (1+x)P_{n-1}^{(\alpha+2,\beta+1)}(x)P_n^{(\alpha+1,\beta)}(x)(1-x)^\alpha(1+x)^\beta dx$$

$$= \frac{\Gamma(n+\alpha+2)}{(n-1)!\Gamma(\alpha+3)} \int_{-1}^1 (1+x)P_n^{(\alpha+1,\beta)}(x)(1-x)^\alpha(1+x)^\beta dx$$

$$= 2\frac{\Gamma(n+\alpha+2)}{(n-1)!\Gamma(\alpha+3)} \int_{-1}^1 P_n^{(\alpha+1,\beta)}(x)(1-x)^\alpha(1+x)^\beta dx.$$

Now, with the same procedure as in (3.37), we get the equation

$$\int_{-1}^{1}(1+x)P_{n-1}^{(\alpha+2,\beta+1)}(x)P_{n}^{(\alpha+1,\beta)}(x)(1-x)^{\alpha}(1+x)^{\beta}dx$$

$$= \frac{\Gamma(n+\alpha+2)}{(n-1)!\Gamma(\alpha+3)}2^{\alpha+\beta+2}\frac{\Gamma(\alpha+1)\Gamma(\beta+n+1)}{\Gamma(\alpha+\beta+n+2)}$$

$$= \frac{2^{\alpha+\beta+2}}{(n-1)!}\frac{\Gamma(\alpha+n+2)\Gamma(\beta+n+1)}{(\alpha+2)(\alpha+1)\Gamma(\alpha+\beta+n+2)}. \tag{3.38}$$

Using the formula (3.31) for the operator $L_{\alpha\beta}$ and formula (3.33) for the derivative of the Jacobi polynomials, we get for the frequency variance of the normalized Christoffel-Darboux kernel:

$$\operatorname{var}_{F}^{\alpha\beta}\left(\tilde{K}_{n}^{(\alpha,\beta)}\right) = \frac{\langle-L_{\alpha\beta}P_{n}^{(\alpha+1,\beta)},P_{n}^{(\alpha+1,\beta)}\rangle_{w_{\alpha\beta}}}{\|P_{n}^{(\alpha+1,\beta)}\|_{w_{\alpha\beta}}^{2}}$$

$$= \frac{\langle-L_{\alpha+1,\beta}P_{n}^{(\alpha+1,\beta)}-(1+\cos t)P_{n}^{(\alpha+1,\beta)'},P_{n}^{(\alpha+1,\beta)}\rangle_{w_{\alpha\beta}}}{\|P_{n}^{(\alpha+1,\beta)}\|_{w_{\alpha\beta}}^{2}}$$

$$= \frac{\langle-L_{\alpha+1,\beta}P_{n}^{(\alpha+1,\beta)},P_{n}^{(\alpha+1,\beta)}\rangle_{w_{\alpha\beta}}}{\|P_{n}^{(\alpha+1,\beta)}\|_{w_{\alpha\beta}}^{2}}$$

$$-\frac{n+\alpha+\beta+2}{2}\frac{\langle(1+\cos t)P_{n-1}^{(\alpha+2,\beta+1)},P_{n}^{(\alpha+1,\beta)}\rangle_{w_{\alpha\beta}}}{\|P_{n}^{(\alpha+1,\beta)}\|_{w_{\alpha\beta}}^{2}}.$$

Now, using formula (3.30) and equation (3.38), we get

$$\operatorname{var}_{F}^{\alpha\beta}\left(\tilde{K}_{n}^{(\alpha,\beta)}\right) = n(n+\alpha+\beta+2)-\frac{1}{\alpha+2}n(n+\alpha+\beta+2)$$

$$= \frac{\alpha+1}{\alpha+2}n(n+\alpha+\beta+2).$$

Finally, for the uncertainty product, we get

$$\operatorname{var}_{S}^{\alpha\beta}\left(\tilde{K}_{n}^{(\alpha,\beta)}\right)\cdot\operatorname{var}_{F}^{\alpha\beta}\left(\tilde{K}_{n}^{(\alpha,\beta)}\right) = \frac{1-\varepsilon_{\alpha\beta}\left(\tilde{K}_{n}^{(\alpha,\beta)}\right)^{2}}{\left|\frac{\alpha-\beta}{\alpha+\beta+2}+\varepsilon_{\alpha\beta}\left(\tilde{K}_{n}^{(\alpha,\beta)}\right)\right|^{2}}\cdot\operatorname{var}_{F}^{\alpha\beta}\left(\tilde{K}_{n}^{(\alpha,\beta)}\right)$$

$$= \frac{(\alpha+\beta+2)^{2}}{\alpha+2}\frac{(2n+\beta+1)(n+\alpha+\beta+2)}{4n}.$$

□

Theorem 3.15 states that also in the more general Jacobi setting the uncertainty product of the normalized Christoffel-Darboux kernel $\tilde{K}_{n}^{(\alpha,\beta)}$ tends linearly to infinity as $n\to\infty$. The Christoffel-Darboux kernel has therefore a much worse space-frequency behavior than the space optimal polynomial $\mathcal{P}_{n}^{(\alpha,\beta)}$ (see Theorem 3.14).

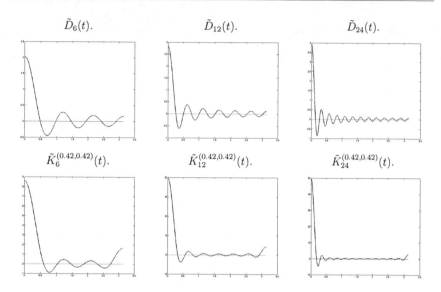

Figure 15: Dirichlet kernel and Christoffel-Darboux kernel for Jacobi expansions with parameters $\alpha = \beta = 0.42$ on $[0, \pi]$.

3.1.4. Space-frequency localization of the de La Vallée Poussin kernel

As a second example, we consider a family of polynomial functions V_n for which the uncertainty product of position and frequency variance tends to the optimal constant $\frac{(\alpha+\beta+2)^2}{4}$ as $n \to \infty$. In this way, we get also an alternative proof for the asymptotic sharpness of the uncertainty principle (1.72) for Jacobi expansions.

The trigonometric polynomial V_n of degree n, known as de La Vallée Poussin kernel (cf. [31, p. 88]), is defined as

$$V_n(t) := (1 + \cos t)^n, \quad n \in \mathbb{N}. \tag{3.39}$$

As in the last section, we denote by $\tilde{V}_n$ the normalized variant of the de La Vallée Poussin kernel, i.e., $\tilde{V}_n(t) = V_n(t)/\|V_n\|_{w_{\alpha\beta}}$. The following Theorem is a slight generalization of [27, Theorem 2.2] proven by Goh and Goodman for ultraspherical expansions.

Theorem 3.16.
For the normalized de La Vallée Poussin kernel $\tilde{V}_n$, the following identities hold:

$$\varepsilon_{\alpha\beta}(\tilde{V}_n) = 1 - \frac{2\alpha + 2}{2n + \alpha + \beta + 2},$$

$$\text{var}_F^{\alpha\beta}(\tilde{V}_n) = \|\tilde{V}_n'\|_{w_{\alpha\beta}}^2 = \frac{(\alpha+1)n^2}{2n+\beta},$$

$$\frac{1 - \varepsilon_{\alpha\beta}(\tilde{V}_n)^2}{|\frac{\alpha-\beta}{\alpha+\beta+2} + \varepsilon_{\alpha\beta}(\tilde{V}_n)|^2} \text{var}_F^{\alpha\beta}(\tilde{V}_n) = \frac{(\alpha+\beta+2)^2}{4}\left(1 + \frac{1}{2n+\beta}\right).$$

Proof. Using the integral formula (1.61), we get the identities

$$\|V_n\|_{w_{\alpha\beta}}^2 = 2^{2n+\alpha+\beta+1} \int_0^\pi \cos^{4n}(\tfrac{t}{2}) \sin^{2\alpha+1}(\tfrac{t}{2}) \cos^{2\beta+1}(\tfrac{t}{2}) dt$$

$$= 2^{2n+\alpha+\beta+1} \frac{\Gamma(\alpha+1)\Gamma(2n+\beta+1)}{\Gamma(\alpha+\beta+2n+2)},$$

$$\|V_n'\|_{w_{\alpha\beta}}^2 = 2^{2n+\alpha+\beta+1} n^2 \int_0^\pi \cos^{4n-2}(\tfrac{t}{2}) \sin^{2\alpha+3}(\tfrac{t}{2}) \cos^{2\beta+1}(\tfrac{t}{2}) dt$$

$$= 2^{2n+\alpha+\beta+1} n^2 \frac{\Gamma(\alpha+2)\Gamma(2n+\beta)}{\Gamma(\alpha+\beta+2n+2)},$$

$$\|V_n\|_{w_{\alpha\beta}}^2 - \varepsilon_{\alpha\beta}(V_n) = 2^{2n+\alpha+\beta+2} \int_0^\pi \sin^2(\tfrac{t}{2}) \cos^{4n}(\tfrac{t}{2}) \sin^{2\alpha+1}(\tfrac{t}{2}) \cos^{2\beta+1}(\tfrac{t}{2}) dt$$

$$= 2^{2n+\alpha+\beta+2} \frac{\Gamma(\alpha+2)\Gamma(2n+\beta+1)}{\Gamma(\alpha+\beta+2n+3)}.$$

Hence, the formulas for $\varepsilon_{\alpha\beta}(\tilde{V}_n)$ and $\text{var}_F^{\alpha\beta}(\tilde{V}_n)$ follow immediately. Moreover, inserting the obtained values for $\varepsilon_{\alpha\beta}(\tilde{V}_n)$ and $\text{var}_F^{\alpha\beta}(\tilde{V}_n)$ in the uncertainty product, a short calculation gives

$$\frac{1 - \varepsilon_{\alpha\beta}(\tilde{V}_n)^2}{|\frac{\alpha-\beta}{\alpha+\beta+2} + \varepsilon_{\alpha\beta}(\tilde{V}_n)|^2} \text{var}_F^{\alpha\beta}(\tilde{V}_n) = \frac{(\alpha+\beta+2)^2}{4}\left(1 + \frac{1}{2n+\beta}\right).$$

$\square$

Hence, although the polynomial $\tilde{V}_n$ is not localized in space as well as the space optimal polynomial $\mathcal{P}_n^{(\alpha,\beta)}$, the de La Vallée Poussin kernel $\tilde{V}_n$ shows a better space-frequency behavior as n tends to infinity. In particular, the frequency variance of $\tilde{V}_n$ increases only linearly in n, whereas $\text{var}_F^{\alpha\beta}(\mathcal{P}_n^{(\alpha,\beta)})$ increases quadratically.

$\tilde{V}_6(t).$ $\qquad\qquad$ $\tilde{V}_{12}(t).$ $\qquad\qquad$ $\tilde{V}_{24}(t).$

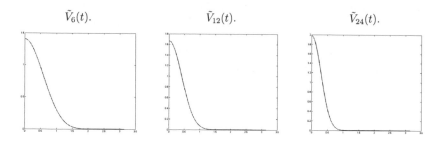

Figure 16: The de La Vallée Poussin kernel normalized in the Chebyshev norm $\|\cdot\|_{w_{\alpha\beta}}$, where $\alpha = \beta = -\frac{1}{2}$.

3.2. Monotonicity of extremal zeros of orthogonal polynomials

To carry the results of Theorem 3.6 over to the setting of a compact two-point homogeneous space, we need an intermediate result concerning the behavior of the extremal zeros of orthogonal polynomials $P_n(x,\tau)$ in terms of a parameter τ. An interesting result in this direction based on the Hellmann-Feynman theorem is due to Ismail [41]. A slightly modified variant of the results in [41] is given by the next theorem.

Theorem 3.17.
Let $Q_n(x,\tau)$, $n \geq 0$, be a family of monic orthogonal polynomials on $[a,b]$ ($-\infty \leq a \leq b \leq \infty$) depending on the parameter τ and fulfilling the three-term recursion formula

$$xQ_n(x,\tau) = Q_{n+1}(x,\tau) + a_n(\tau)Q_n(x,\tau) + b_n(\tau)Q_{n-1}(x,\tau), \quad n \geq 0, \qquad (3.40)$$
$$Q_{-1}(x,\tau) = 0, \quad Q_0(x,\tau) = 1.$$

Assume that the coefficients $a_n(\tau)$ ($n \geq 0$) and $b_n(\tau) > 0$ ($n \geq 1$) are differentiable monotone decreasing (increasing) functions of the parameter τ. Then the largest zero of the polynomial $Q_n(x,\tau)$ is also a differentiable monotone decreasing (increasing) function of the parameter τ.
On the other hand, if the coefficients $b_n(\tau)$ are monotone decreasing (increasing) and the coefficients $a_n(\tau)$ are monotone increasing (decreasing) functions, then the smallest zero of $Q_n(x,\tau)$ is differentiable monotone increasing (decreasing). If one of the coefficients $a_k(\tau)$ or $b_k(\tau)$, $k \leq n$, is strictly monotone decreasing or increasing, then, in the above statement, the smallest and the largest zero of $Q_n(x,\tau)$ are also strictly monotone.

Proof. Let $\lambda(\tau)$ be the largest zero of $Q_n(x,\tau)$. Clearly, all zeros of $Q_n(x,\tau)$ are differentiable functions of τ. Then, the Hellmann-Feynman theorem (see Theorem 7.3.1 and, in particular, equation (7.3.8) in [42]) combined with the three-term recurrence formula

(3.40) of the polynomials $Q_n(x, \tau)$ implies the formula

$$\left(\sum_{k=0}^{n-1} \frac{Q_k^2(\lambda, \tau)}{\zeta_k} \right) \frac{d\lambda(\tau)}{d\tau} = \sum_{k=0}^{n-1} \frac{Q_k(\lambda, \tau)}{\zeta_k} \left(a_k'(\tau) Q_k(\lambda, \tau) + b_k'(\tau) Q_{k-1}(\lambda, \tau) \right) \qquad (3.41)$$

where a_k' and b_k' denote differentiation with respect to τ and $\zeta_k = \prod_{i=1}^{k} b_i(\tau)$. Since the polynomials $Q_k(x, \tau)$ are monic, we have $Q_k(x, \tau) = \prod_{i=1}^{k}(x - x_i(\tau))$, where $x_i^k(\tau)$, $i = 1, \ldots, k$ denote the k distinct real zeros of $Q_k(x, \tau)$ in (a, b). Hence, $Q_k(b, \tau) > 0$. Moreover, since $\lambda(\tau)$ is the largest zero of $Q_n(x, \tau)$, we have due to the interlacing property of the polynomials $Q_k(x, \tau)$ (see [83, Theorem 3.3.2])) that $Q_k(\lambda, \tau) > 0$ for $k = 0, \ldots, n - 1$. Therefore, if $a_k(\tau)$ and $b_k(\tau)$ are decreasing (increasing) functions of the parameter τ, then the right hand side of equation (3.41) is negative (positive) and the first statement of the Theorem is shown. A similar argumentation for the smallest zero (keeping in mind that $\text{sign}(Q_k(a, \tau)) = (-1)^k$) implies the second statement. The statement for the strict monotonicity follows directly from formula (3.41). $\qquad \square$

Now, we will use Theorem 3.17 to prove that the largest zero of the associated Jacobi and ultraspherical polynomials is decreasing if certain parameters are increased.

Corollary 3.18.
Let $c \geq 0$ (if $c > 0$, assume that $2c + \alpha + \beta > 0$), $\alpha \geq 0$, $\beta \geq -1/2$ and $\beta \leq \max\{\frac{1}{2}, 2\alpha\}$. Then, the largest zero $\lambda(\alpha)$ of the associated Jacobi polynomial $p_n^{(\alpha, \beta)}(x, c)$ is a decreasing function of the parameter α.

Proof. We consider the monic associated Jacobi polynomials $Q_n^{(\alpha, \beta)}(x, c)$ as orthogonal polynomials depending on the parameter α. The polynomials $Q_n^{(\alpha, \beta)}(x, c)$ are defined by the three-term recurrence relation

$$x Q_n^{(\alpha, \beta)}(x, c) = Q_{n+1}^{(\alpha, \beta)}(x, c) + a_{n+c}(\alpha) Q_n^{(\alpha, \beta)}(x, c) + b_{n+c}^2(\alpha) Q_{n-1}^{(\alpha, \beta)}(x, c), \quad n \geq 0, \quad (3.42)$$
$$Q_{-1}^{(\alpha, \beta)}(x, c) = 0, \quad Q_0^{(\alpha, \beta)}(x, c) = 1,$$

where the coefficients $a_{n+c}(\alpha)$ and $b_{n+c}^2(\alpha)$ are given by (see Table 1.1 in [25] for the coefficients of the Jacobi polynomials)

$$a_{n+c}(\alpha) = \frac{\beta^2 - \alpha^2}{(2n + 2c + \alpha + \beta)(2n + 2c + 2 + \alpha + \beta)}, \quad n \geq 0,$$

$$b_{n+c}^2(\alpha) = \frac{4(n + c)(n + c + \alpha)(n + c + \beta)(n + c + \alpha + \beta)}{(2n + 2c + \alpha + \beta)^2(2n + 2c + \alpha + \beta + 1)(2n + 2c + \alpha + \beta - 1)}, \quad n \geq 1.$$

Now, we have to check that the assumptions of Theorem 3.17 hold. In particular, we show that $a_{n+c}(\alpha)$ and $b_{n+c}^2(\alpha)$ are decreasing functions of the variable α.

First, we consider the derivative a'_{n+c}. For $n \geq 0$, $c > 0$, we have

$$
\begin{aligned}
a'_{n+c}(\alpha) &= \frac{\left(-2\alpha - \frac{\beta^2-\alpha^2}{(2n+2c+\alpha+\beta)} - \frac{\beta^2-\alpha^2}{(2n+2c+2+\alpha+\beta)}\right)}{(2n+2c+\alpha+\beta)(2n+2c+2+\alpha+\beta)} \\
&= -2\frac{\left(4\alpha(n+c)^2 + 2(n+c)(2\alpha+(\alpha+\beta)^2) + (1+\beta)(\alpha+\beta)^2\right)}{(2n+2c+\alpha+\beta)^2(2n+2c+2+\alpha+\beta)^2}.
\end{aligned}
$$

Since we assumed that $\alpha \geq 0$ and $\beta \geq -1/2$, the term on the right hand side is always negative. It remains to check the case $n = 0$, $c = 0$. In this case, we get

$$
a'_0(\alpha) = -\frac{2(1+\beta)}{(\alpha+\beta+2)^2} < 0.
$$

Thus, $a_{n+c}(\alpha)$ is a monotone decreasing function of the parameter α if $\alpha \geq 0$, $\beta \geq -\frac{1}{2}$ and $n \geq 0$.

Next, we examine the derivative $(b^2_{n+c})'(\alpha)$. For $n \geq 1$, we get

$$
\begin{aligned}
(b^2_{n+c})'(\alpha) = b^2_{n+c}(\alpha) &\left(\frac{1}{n+c+\alpha} + \frac{1}{n+c+\alpha+\beta} - \frac{2}{2n+2c+\alpha+\beta}\right. \\
&\left. - \frac{1}{2n+2c+\alpha+\beta+1} - \frac{1}{2n+2c+\alpha+\beta-1}\right).
\end{aligned}
$$

We consider first the case when $\alpha \geq \frac{1}{2}$ and $-\frac{1}{2} \leq \beta \leq 2\alpha$. Here, we get the upper bound

$$
\begin{aligned}
(b^2_{n+c})'(\alpha) &\leq b^2_{n+c}(\alpha)\left(\frac{1}{n+c+\alpha} + \frac{1}{n+c+\alpha+\beta} - \frac{4}{2n+2c+\alpha+\beta}\right) \\
&= b^2_{n+c}(\alpha)\frac{-2(n+c)\alpha + (\beta+\alpha)(\beta-2\alpha)}{(n+c+\alpha)(n+c+\alpha+\beta)(2n+2c+\alpha+\beta)} \leq 0.
\end{aligned}
$$

Hence $(b^2_{n+c})'(\alpha)$ is negative if $\alpha \geq \frac{1}{2}$, $-\frac{1}{2} \leq \beta \leq 2\alpha$, $c \geq 0$ and $n \geq 1$. Next, we consider the case $0 \leq \alpha \leq \frac{1}{2}$, $-\frac{1}{2} \leq \beta \leq \frac{1}{2}$ and $n \geq 1$. In this case, we get the estimate

$$
(b_{n+c}^2)'(\alpha) = b_{n+c}^2(\alpha) \left(\frac{2}{(n + c + \alpha + \frac{\beta}{2}) - \frac{\beta^2}{4n+4c+4\alpha+2\beta}} \right.
$$

$$
\left. - \frac{2}{2n + 2c + \alpha + \beta} - \frac{2}{(2n + 2c + \alpha + \beta) - \frac{1}{2n+2c+\alpha+\beta}} \right)
$$

$$
= b_{n+c}^2(\alpha) \left(\frac{2}{(n + c + \alpha + \frac{\beta}{2}) - \frac{\beta^2}{4n+4c+4\alpha+2\beta}} \right.
$$

$$
- \frac{2}{\left(2n + 2c + \alpha + \beta - \frac{1}{4n+4c+2\alpha+2\beta}\right) + \frac{1}{4n+4c+2\alpha+2\beta}}
$$

$$
\left. - \frac{2}{\left(2n + 2c + \alpha + \beta - \frac{1}{4n+4c+2\alpha+2\beta}\right) - \frac{1}{4n+4c+2\alpha+2\beta}} \right)
$$

$$
\leq b_{n+c}^2(\alpha) \left(\frac{2}{n + c + \alpha + \frac{\beta}{2} - \frac{\beta^2}{4n+4c+4\alpha+2\beta}} - \frac{2}{n + c + \frac{\alpha}{2} + \frac{\beta}{2} - \frac{1}{8n+8c+4\alpha+4\beta}} \right).
$$

Since $(n + c + \alpha + \frac{\beta}{2}) \geq (n + c + \frac{\alpha}{2} + \frac{\beta}{2})$ and $\frac{\beta^2}{4n+4c+4\alpha+2\beta} \leq \frac{1}{8n+8c+4\alpha+4\beta}$, we can see that also in this case the derivative $(b_{n+c}^2)'(\alpha) \leq 0$ is negative.

In total, we can conclude that $(b_{n+c}^2)'(\alpha) \leq 0$ and that $b_{n+c}^2(\alpha)$ is a monotone decaying function of the parameter α if $n \geq 1$, $c \geq 0$, $\alpha \geq 0$ and $-\frac{1}{2} \leq \beta \leq \max\{\frac{1}{2}, 2\alpha\}$. By Theorem 3.17, the largest zero $\lambda(\alpha)$ of $Q_n^{(\alpha,\beta)}(x,c)$ is therefore a decreasing function of the parameter α. Since the polynomials $p_n^{(\alpha,\beta)}(x,c)$ are given by $p_n^{(\alpha,\beta)}(x,c) = b_{1+c} \cdots b_{n+c} Q_n^{(\alpha,\beta)}(x,c)$ (compare the recurrence relations (3.42) and (3.12)), the same statement holds also for the polynomials $p_n^{(\alpha,\beta)}(x,c)$. $\qquad \square$

Corollary 3.19.
If $\alpha \geq 0$, then the largest zero $\lambda(\alpha)$ of the associated ultraspherical polynomials $p_n^{(\alpha,\alpha)}(x,c)$, $c \geq 0$, is a decreasing function of the parameter α.

Proof. By (3.42), the monic associated ultraspherical polynomials $Q_n^{(\alpha,\alpha)}(x,c)$ satisfy the three-term recurrence relation

$$
x Q_n^{(\alpha,\alpha)}(x,c) = Q_{n+1}^{(\alpha,\alpha)}(x,c) + b_{n+c}^2(\alpha) Q_{n-1}^{(\alpha,\alpha)}(x,c), \quad n \geq 0,
$$

$$
Q_{-1}^{(\alpha,\alpha)}(x,c) = 0, \quad Q_0^{(\alpha,\alpha)}(x,c) = 1,
$$

with the coefficients

$$
a_{n+c}(\alpha) = 0, \quad n \geq 0,
$$

$$
b_{n+c}^2(\alpha) = \frac{(n+c)(n+c+2\alpha)}{4\left(n + c + \alpha + \frac{1}{2}\right)\left(n + c + \alpha - \frac{1}{2}\right)}, \quad n \geq 1.
$$

Thus, we have $a'_{n+c}(\alpha) = 0$, and for $(b^2_{n+c})'(\alpha)$, $\alpha \geq 0$, we get

$$(b^2_{n+c})'(\alpha) = b^2_{n+c}(\alpha) \left(\frac{2}{n+c+2\alpha} - \frac{1}{n+c+\alpha+\frac{1}{2}} - \frac{1}{n+c+\alpha-\frac{1}{2}} \right)$$

$$= b^2_{n+c}(\alpha) \frac{-2\alpha(n+c-1) - 2(\alpha+\frac{1}{2})^2}{(n+c+2\alpha)(n+c+\alpha+\frac{1}{2})(n+c+\alpha-\frac{1}{2})} \leq 0.$$

Thus, $b^2_{n+c}(\alpha)$ is a decreasing function of the parameter α if $\alpha \geq 0$. Due to Theorem 3.17, the largest zero of $Q_n^{(\alpha,\alpha)}(x,c)$ is a decreasing function of the parameter α, and so is the largest zero $\lambda(\alpha)$ of $p_n^{(\alpha,\alpha)}(x,c) = b_{1+c} \cdots b_{n+c} Q_n^{(\alpha,\alpha)}(x,c)$. $\qquad\square$

Also for the next result, we can use Theorem 3.17.

Corollary 3.20.
Let $\alpha \geq \beta$, $\alpha + \beta \geq 0$, $c > 0$, $2c + \alpha + \beta > 0$ and $0 \leq \tau \leq c$. Then, the largest zero $\lambda(\tau)$ of the associated Jacobi polynomial $p_n^{(\alpha+\tau,\beta+\tau)}(x, c-\tau)$, $c \geq 0$, is a strictly decreasing function of the parameter τ. Similarly, if $\alpha \geq 0$, $\beta > -1$ and $0 \leq \sigma \leq c$, then the largest zero $\lambda(\sigma)$ of the associated Jacobi polynomial $p_n^{(\alpha+2\sigma,\beta)}(x, c-\sigma)$ is a strictly decreasing function of the parameter σ.

Proof. First, we consider the monic associated Jacobi polynomials $Q_n^{(\alpha+\tau,\beta+\tau)}(x, c-\tau)$ as a family of orthogonal polynomials depending on the parameter τ. In this case, the coefficients $a_n(\tau)$ and $b^2_n(\tau)$ in the three-term recurrence formula (3.42) are given by

$$a_n(\tau) = \frac{\beta^2 - \alpha^2 + 2\tau(\beta - \alpha)}{(2n+2c+\alpha+\beta)(2n+2c+2+\alpha+\beta)}, \quad n \geq 0, \tag{3.43}$$

$$b^2_n(\tau) = \frac{4(n+c-\tau)(n+c+\alpha)(n+c+\beta)(n+c+\alpha+\beta+\tau)}{(2n+2c+\alpha+\beta)^2(2n+2c+\alpha+\beta+1)(2n+2c+\alpha+\beta-1)}, \quad n \geq 1. \tag{3.44}$$

For the derivatives, we get

$$a'_n(\tau) = \frac{2(\beta-\alpha)}{(2n+2c+\alpha+\beta)(2n+2c+2+\alpha+\beta)} \leq 0 \quad \text{if } \alpha \geq \beta, \tag{3.45}$$

$$(b^2_n)'(\tau) = b^2_n(\tau) \left(\frac{1}{n+c+\alpha+\beta+\tau} - \frac{1}{n+c-\tau} \right) \leq 0 \quad \text{if } \alpha+\beta \geq 0. \tag{3.46}$$

Further, inequality (3.46) is strict if $\tau > 0$. Thus, by Theorem 3.17, the largest zero of $Q_n^{(\alpha+\tau,\alpha+\tau)}(x, c-\tau)$, and therefore also the largest zero of the normalized polynomial $p_n^{(\alpha+\tau,\alpha+\tau)}(x, c-\tau)$, is a strictly decreasing function of the parameter $0 \leq \tau \leq c$.

Now, we consider the monic associated polynomials $Q_n^{(\alpha+2\sigma,\beta)}(x, c-\sigma)$ depending on the parameter σ. The coefficients of the three-term recurrence formula are given by

$$a_n(\sigma) = \frac{\beta^2 - (\alpha + 2\sigma)^2}{(2n + 2c + \alpha + \beta)(2n + 2c + 2 + \alpha + \beta)}, \quad n \geq 0,$$

$$b_n^2(\sigma) = \frac{4(n + c - \sigma)(n + c + \alpha + \sigma)(n + c - \sigma + \beta)(n + c + \alpha + \sigma + \beta)}{(2n + 2c + \alpha + \beta)^2(2n + 2c + \alpha + \beta + 1)(2n + 2c + \alpha + \beta - 1)}, \quad n \geq 1.$$

In this case, we get for the derivatives

$$a_n'(\sigma) = \frac{-4(\alpha + 2\sigma)}{(2n + 2c + \alpha + \beta)(2n + 2c + 2 + \alpha + \beta)},$$

$$(b_n^2)'(\sigma) = b_n^2(\sigma)\left(-\frac{1}{n + c - \sigma} + \frac{1}{n + c + \alpha + \sigma} - \frac{1}{n + c + \beta - \sigma} + \frac{1}{n + c + \alpha + \beta + \sigma}\right)$$

$$= -b_n^2(\sigma)\left(\frac{\alpha + 2\sigma}{(n + c - \sigma)(n + c + \alpha + \sigma)} + \frac{\alpha + 2\sigma}{(n + c + \alpha + \beta + \sigma)(n + c + \beta - \sigma)}\right).$$

Both, $a_n'(\sigma)$ and $(b_n^2)'(\sigma)$ are negative if $\alpha \geq 0$ and strictly negative if $\tau > 0$. Thus, the largest zero of the polynomial $Q_n^{(\alpha+2\sigma,\beta)}(x, c - \sigma)$ and of the normalized polynomial $p_n^{(\alpha+2\sigma,\beta)}(x, c - \sigma)$ is a strictly decreasing function of the parameter $0 \leq \sigma \leq c$. $\qquad\square$

If $\alpha = 0$ and $\beta < 0$, it is not possible to use Corollary 3.20 to prove that the largest zero $\lambda(\tau)$ of the polynomial $p_n^{(\tau,\beta+\tau)}(x, c - \tau)$ is a decreasing function of the parameter τ. Nevertheless, we can show the following result.

Theorem 3.21.
Let $\alpha \geq 0$, $-1 < \beta \leq 0$, $\alpha \leq |\beta|$ and $2c + \alpha + \beta > 0$. Then all the zeros of the associated Jacobi polynomial $p_n^{(\alpha,\beta)}(x, c)$ are larger than the respective zeros of the polynomial $p_n^{(-\beta,-\alpha)}(x, c + \alpha + \beta)$.

Proof. As in Corollary 3.20, we consider the associated Jacobi polynomials $P_{n+1}(\tau)(x) = p_{n+1}^{(\alpha+\tau,\beta+\tau)}(x, c - \tau)$ with the Jacobi matrix

$$\mathbf{J}_n(\tau) = \begin{pmatrix} a_0(\tau) & b_1(\tau) & 0 & 0 & \cdots & 0 \\ b_1(\tau) & a_1(\tau) & b_2(\tau) & 0 & \cdots & 0 \\ 0 & b_2(\tau) & a_2(\tau) & b_3(\tau) & \ddots & \vdots \\ \vdots & \ddots & \ddots & \ddots & \ddots & 0 \\ 0 & \cdots & 0 & b_{n-2}(\tau) & a_{n-1}(\tau) & b_{n-1}(\tau) \\ 0 & \cdots & \cdots & 0 & b_{n-1}(\tau) & a_n(\tau) \end{pmatrix},$$

and the coefficients $a_n(\tau)$ and $b_n(\tau) = \sqrt{b_n^2(\tau)}$ given in (3.43) and (3.44). Then, $P_{n+1}(0)(x) = p_{n+1}^{(\alpha,\beta)}(x, c)$ and $P_{n+1}(-\alpha - \beta)(x) = p_{n+1}^{(-\beta,-\alpha)}(x, c + \alpha + \beta)$. Now, the zeros of the polynomials $P_{n+1}(0)$ and $P_{n+1}(-\alpha - \beta)$ correspond to the eigenvalues of the Jacobi matrices $\mathbf{J}_n(0)$ and $\mathbf{J}_n(-\alpha - \beta)$, respectively. Since

$$\mathbf{J}_n(-\alpha - \beta) - \mathbf{J}(0) = \operatorname{diag}\big(a_0(-\alpha - \beta) - a_0(0), \cdots, a_n(-\alpha - \beta) - a_n(0)\big)$$

is a diagonal matrix with the negative entries

$$a_k(-\alpha - \beta) - a_k(0) = 2\frac{\alpha^2 - \beta^2}{(2k + 2c + \alpha + \beta)(2k + 2c + \alpha + \beta + 2)},$$

the eigenvalues of $\mathbf{J}_n(0)$ are larger than the eigenvalues of $\mathbf{J}_n(-\alpha - \beta)$. Thus, the same holds for the zeros of the polynomials $p_{n+1}^{(\alpha,\beta)}(x,c)$ and $p_{n+1}^{(-\beta,-\alpha)}(x,c + \alpha + \beta)$. $\square$

3.3. Optimally space localized spherical polynomials on compact two-point homogeneous spaces

3.3.1. Compact two-point homogeneous spaces

A connected Riemannian manifold M is called two-point homogeneous if for any two pairs of points p_1, p_2 and q_1, q_2 on M with $d(p_1, p_2) = d(q_1, q_2)$ there exists an isometry I on M carrying p_1 to q_1 and p_2 to q_2. According to Wang [84], the compact two-point homogeneous spaces are precisely the spheres and the projective spaces introduced in Section 2.6.1 and 2.6.2, and can be listed as follows (see also [1, Section 3], [24, p. 176-177] and [35, p. 170]):

(i) The sphere $\mathbb{S}_r^d$, $r > 0$, $d = 1, 2, 3, \ldots$

(ii) The real projective space $\mathbb{RP}_r^d$, $r > 0$, $d = 2, 3, 4, \ldots$

(iii) The complex projective space $\mathbb{CP}_r^d$, $r > 0$, $d = 4, 6, 8, \ldots$

(iv) The quaternionic projective space $\mathbb{HP}_r^d$, $r > 0$, $d = 8, 12, 16, \ldots$

(v) The Cayley plane $\mathbb{Ca}_r$,

where the superscripts d denote the real dimension of the respective manifold. In the literature, these spaces are also known as the compact symmetric spaces of rank one (see [34, Section IX.5]). From now on, the symbol M will denote a compact two-point homogeneous space and p a point on M.

The compact two-point homogeneous spaces are very similar in their geometry. In particular, these spaces have the remarkable property that all their geodesics are closed curves of length $2r\pi$ (see [36, VII, Proposition 10.2]). Further, the diameter of M is given by

$$\operatorname{diam}(M) = \sup\{d(q_1, q_2) : q_1, q_2 \in M\} = r\pi$$

and the cut locus C_p of p, with the Riemannian structure induced by M, is itself a compact two-point homogeneous space (see [36, VII, Proposition 10.4]). For the details on isometry groups, symmetric spaces and, in particular, compact two-point homogeneous spaces, we refer to the classics [34], [36] and [37].

If G denotes the maximal connected group of isometries of M and $K = \{I \in G : Ip = p\}$, then M corresponds to the homogeneous space G/K. The isometry groups G and K of each two-point homogeneous space M are well-known and are listed, for instance, in [24, p. 177].

The Hilbert space $L^2(M)$ of square integrable functions on M can be decomposed into a direct orthogonal sum of finite-dimensional G-invariant, G-irreducible subspaces $\mathcal{H}_l^M$, such that (see [37, Chapter V, Theorem 4.3, and Chapter II, Proposition 4.11])

$$L^2(M) = \bigoplus_{l=0}^{\infty} \mathcal{H}_l^M.$$

The G-invariant subspace $\mathcal{H}_l^M$ corresponds to the eigenspace of the Laplace-Beltrami operator Δ_M with respect to the eigenvalue $l(l + \frac{d}{2} + \beta_M)$, i.e.,

$$\mathcal{H}_l^M = \{f \in C^{\infty}(M) : \Delta_M f = l(l + \tfrac{d}{2} + \beta_M)f\}, \tag{3.47}$$

where the parameter β_M is given as $\beta_M = \frac{d-2}{2}$ in the case that M is the sphere $\mathbb{S}_r^d$ and as $\beta_M = -\frac{1}{2}, 0, 1, 3$ in the case that M is one of the projective spaces $\mathbb{RP}_r^d$, $\mathbb{CP}_r^d$, $\mathbb{HP}_r^d$ and $\mathbb{C}a_r$, respectively. The dimensions $\dim \mathcal{H}_l^M$ of the spaces $\mathcal{H}_l^M$ can be computed explicitly and are collected in Table 1.

Table 1: The dimension of the subspaces $\mathcal{H}_l^M$ ([80], p. 90).

Space	$\dim \mathcal{H}_l^M$
$\mathbb{S}_r^d$	$\dfrac{(2l + d - 1)(l + d - 2)!}{l!(d-1)!}$
$\mathbb{RP}_r^d$	$\dfrac{(4l + d - 1)(2l + d - 2)!}{(2l)!(d-1)!}$
$\mathbb{CP}_r^d$	$\dfrac{(2l + \frac{d}{2})(l + \frac{d}{2} - 1)!(l + \frac{d}{2} - 1)!}{(\frac{d}{2})!(\frac{d}{2} - 1)!l!l!}$
$\mathbb{HP}_r^d$	$\dfrac{(2l + \frac{d}{2} + 1)(l + \frac{d}{2})!(l + \frac{d}{2} - 1)!}{(\frac{d}{2} + 1)!(\frac{d}{2} - 1)!(l + 1)!l!}$
$\mathbb{C}a_r$	$\dfrac{(2l + 11)(l + 10)!(l + 7)!3!}{11!l!(l + 3)!7!}$

In geodesic polar coordinates (t, ξ) at a point p, the weight functions Θ_M are given as (see equations (2.148) and (2.160) or [35, p. 169])

$$\Theta_M(t, \xi) = (2r)^{d-1} \sin^{d-1}(\tfrac{t}{2r}) \cos^{2\beta_M+1}(\tfrac{t}{2r}), \tag{3.48}$$

and the radial part $\Delta_{p,t}^M$ of the Laplace-Beltrami operator Δ_M reads as

$$(\Delta_{p,t}^M f)^*(t, \xi) = \frac{\partial^2}{\partial t^2} f^*(t, \xi) + \frac{d - 2 - 2\beta_M + (d + 2\beta_M) \cos(\tfrac{t}{r})}{2r \sin(\tfrac{t}{r})} \frac{\partial}{\partial t} f^*(t, \xi). \tag{3.49}$$

Moreover, each subspace $\mathcal{H}_l^M$ contains only one radial function P_l^M that depends solely on the distance $t = d(q, p)$ to the point p and that is normalized such that $\|P_l^M\|_M = 1$, $P_l^M(p) > 0$. In terms of Jacobi polynomials, the radial functions P_l^M can be written as (see [1, p. 131-132], [24, p. 178] and [37, V, Theorem 4.5])

$$P_l^{M*}(t) = \frac{1}{|\mathfrak{S}_p|^{\frac{1}{2}} 2^{\frac{d-2-2\beta_M}{4}} r^{\frac{d}{2}}} p_l^{(\frac{d-2}{2}, \beta_M)}(\cos(\tfrac{t}{r})), \quad t \in [0, r\pi], \tag{3.50}$$

where $|\mathfrak{S}_p|$ denotes the volume of the $(d-1)$-dimensional unit sphere $\mathfrak{S}_p$.

Next, we are giving an orthonormal basis for the subspaces $\mathcal{H}_l^M$. To this end, we need an orthonormal basis for the subspaces $\mathcal{H}_k^{\mathfrak{S}_p} \subset L^2(\mathfrak{S}_p)$ on the unit sphere $\mathfrak{S}_p$. Such a basis is given by the orthonormal spherical harmonics $Y_{k,j}^{d-1}$, $1 \leq j \leq \dim \mathcal{H}_k^{\mathfrak{S}_p}$, of order k in $d-1$ dimensions (for the details, see [60]). For $d = 2$, the spherical harmonics $Y_{k,j}^1$, $j \in \{1, 2\}$, are identified with the characters $\frac{1}{\sqrt{2\pi}} e^{i(-1)^j kt}$, $k \in \mathbb{N}$, on $[-\pi, \pi]$ and the space $\mathcal{H}_k^{\mathfrak{S}_p}$ with span $\left\{ \frac{1}{\sqrt{2\pi}} e^{ikt}, \frac{1}{\sqrt{2\pi}} e^{-ikt} \right\}$. In the following, we use the symbol $N(d-1, k)$ as an abbreviation for the dimension of the spaces $\mathcal{H}_k^{\mathfrak{S}_p}$.

Proposition 3.22.
If $M = \mathbb{S}_r^d$, $d \geq 2$, an orthonormal basis for the space $\mathcal{H}_l^{\mathbb{S}_r^d}$ is given in geodesic polar coordinates at $p \in \mathbb{S}_r^d$ by the functions

$$P_{l,k,j}^{\mathbb{S}_r^d *}(t, \xi) = r^{-\frac{d}{2}} \sin^k(\tfrac{t}{r}) p_{l-k}^{(\frac{d-2}{2}+k, \frac{d-2}{2}+k)}(\cos(\tfrac{t}{r})) Y_{k,j}^{d-1}(\xi), \tag{3.51}$$

$$0 \leq k \leq l, \quad 1 \leq j \leq N(d-1, k).$$

If $M = \mathbb{RP}_r^d$, $d \geq 2$, an orthonormal basis for $\mathcal{H}_l^{\mathbb{RP}_r^d}$ is given by

$$P_{l,k,j}^{\mathbb{RP}_r^d *}(t, \xi) = 2 P_{2l,k,j}^{\mathbb{S}_{2r}^d *}(t, \xi) = 2(2r)^{-\frac{d}{2}} \sin^k(\tfrac{t}{2r}) p_{2l-k}^{(\frac{d-2}{2}+k, \frac{d-2}{2}+k)}(\cos(\tfrac{t}{2r})) Y_{k,j}^{d-1}(\xi), \tag{3.52}$$

$$= \begin{cases} 2^{\frac{-d+2k+1}{4}} r^{-\frac{d}{2}} \sin^k(\tfrac{t}{2r}) p_{l-\frac{k}{2}}^{(\frac{d-2}{2}+k, -\frac{1}{2})}(\cos(\tfrac{t}{r})) Y_{k,j}^{d-1}(\xi), & k \text{ even}, \\ 2^{\frac{-d+2k+3}{4}} r^{-\frac{d}{2}} \sin^k(\tfrac{t}{2r}) \cos(\tfrac{t}{2r}) p_{l-\frac{k+1}{2}}^{(\frac{d-2}{2}+k, \frac{1}{2})}(\cos(\tfrac{t}{r})) Y_{k,j}^{d-1}(\xi), & k \text{ odd}, \end{cases}$$

$$0 \leq k \leq 2l, \quad 0 \leq j \leq N(d-1, k).$$

Finally, if M is one of the projective spaces $\mathbb{CP}_r^d$, $d \geq 4$, or $\mathbb{HP}_r^d$, $d \geq 8$, or $\mathbb{Ca}_r$, then an explicit orthonormal basis for $\mathcal{H}_l^M$ is given by the functions

$$P_{l,k,j}^{M*}(t,\xi) = 2^{\frac{-d+2+2\beta_M+4k}{4}} r^{-\frac{d}{2}} \sin^{2k}(\tfrac{t}{2r}) p_{l-k}^{(\frac{d-2}{2}+2k,\beta_M)}(\cos(\tfrac{t}{r})) Y_{k,j}^{d-1}(\xi), \tag{3.53}$$

$$0 \leq k \leq l, \quad 0 \leq j \leq N(d-1,k).$$

Proof. We will give no explicit proofs at this place, but refer to the literature. The proof that the spherical harmonics $P_{l,k,j}^{\mathbb{S}^d*}(t,\xi)$ and $Y_{k,j}^{d-1}(\xi)$ form an orthonormal basis of $\mathcal{H}_l^{\mathbb{S}^d}$ and $\mathcal{H}_k^{\mathbb{S}_r}$, respectively, can be found in the books [60] and [14]. The assertion for an arbitrary sphere $\mathbb{S}_r^d$ follows by scaling with r. The assertion for the real projective space is an immediate consequence of the fact that $\mathbb{RP}_r^d$ is the quotient of $\mathbb{S}_{2r}^d$ under the antipodal map $A : x \to -x$ and that $\mathcal{H}_l^{\mathbb{RP}_r^d}$ can be identified with $\mathcal{H}_{2l}^{\mathbb{S}_{2r}^d}$. The second identity in (3.52) follows from [83, Theorem 4.1]. The proof for the remaining projective spaces can be found in [80]. Hereby, the formula in (3.53) follows from [80, Theorem 4.22] with the coordinate change $t \to \sin(\tfrac{t}{2r})^2$. A related proof can also be found in the article [46]. $\square$

In Proposition 3.22, the functions $P_{l,0,1}^M$ correspond to the radial functions P_l^M defined in (3.50). Now, as in (2.91), we introduce the generalized mean value of a function $f \in L^2(M)$ at a point $p \in M$ by

$$\varepsilon_p^M(f) = \int_{\mathfrak{S}_p} \int_0^{r\pi} \cos(\tfrac{t}{r}) |f(t,\xi)|^2 \Theta_M(t) dt d\mu(\xi). \tag{3.54}$$

Then, by Corollary 2.60 and Corollary 2.61, we know that the following uncertainty inequality holds for all normalized functions $f \in L^2(M) \cap \mathcal{D}(\frac{\partial}{\partial t_*}; M)$ satisfying $\varepsilon_p^M(f) \neq \frac{2\beta_M+2-d}{2\beta_M+2+d}$:

$$\mathrm{var}_{S,p}^M(f) \cdot \mathrm{var}_{F,p}^M(f) > \frac{d^2}{4}, \tag{3.55}$$

where

$$\mathrm{var}_{S,p}^M(f) = r^2 \frac{1 - \varepsilon_p^M(f)^2}{\left(\frac{d-2-2\beta_M}{2d} + \frac{d+2+2\beta_M}{2d}\varepsilon_p^M(f)\right)^2}, \tag{3.56}$$

$$\mathrm{var}_{F,p}^M(f) = \left\| \frac{\partial}{\partial t_*} f \right\|_M^2. \tag{3.57}$$

Moreover, we know that the constant $\frac{d^2}{4}$ on the right hand side of (3.55) is optimal.

3.3.2. Optimally space localized spherical polynomials

According to the polynomial subspaces of $L^2([0,\pi], w_{\alpha\beta})$ introduced in Definition 3.1, we define now finite-dimensional subspaces of the Hilbert space $L^2(M)$ spanned by the basis functions of the spaces $\mathcal{H}_l^M$.

Definition 3.23. As subspaces of $L^2(M)$, we consider:

(1) The space spanned by the basis functions of $\mathcal{H}_l^M$, $0 \le l \le n$:

$$\Pi_n^M := \left\{ P: \ P^*(t,\xi) = \sum_{l=0}^{n} \sum_{k=0}^{l} \sum_{j=1}^{N(d-1,k)} c_{l,k,j} P_{l,k,j}^{M*}(t,\xi), \ c_{l,k,j} \in \mathbb{C} \right\}. \qquad (3.58)$$

(2) The space spanned by the basis functions of $\mathcal{H}_l^M$, $m \le l \le n$:

$$\Pi_{m,n}^M := \left\{ P: \ P^*(t,\xi) = \sum_{l=m}^{n} \sum_{k=0}^{l} \sum_{j=1}^{N(d-1,k)} c_{l,k,j} P_{l,k,j}^{M*}(t,\xi), \ c_{l,k,j} \in \mathbb{C} \right\}. \qquad (3.59)$$

(3) The space spanned by a radial function $\mathcal{R}$ given by $\mathcal{R}^*(t) = P_{m-1}^{M*}(t) + \sum_{l=0}^{m-1} e_l P_l^{M*}(t)$ and the basis functions of $\mathcal{H}_l^M$, $m \le l \le n$:

$$\Pi_{\mathcal{R},n}^M := \left\{ P: \ P^*(t,\xi) = c_{m-1}\mathcal{R}^*(t) + \sum_{l=m}^{n} \sum_{k=0}^{l} \sum_{j=1}^{N(d-1,k)} c_{l,k,j} P_{l,k,j}^{M*}(t,\xi), \ c_{l,k,j}, \in \mathbb{C} \right\}. \qquad (3.60)$$

The functions in the space Π_n^M are referred to as spherical polynomials of degree less than n and the functions in $\Pi_{m,n}^M$ as spherical wavelets. The unit spheres in the spaces Π_n^M, $\Pi_{m,n}^M$ and $\Pi_{\mathcal{R},n}^M$ are defined as

$$\mathbb{S}_n^M := \left\{ P \in \Pi_n^M : \ \|P\|_M = 1 \right\},$$
$$\mathbb{S}_{m,n}^M := \left\{ P \in \Pi_{m,n}^M : \ \|P\|_M = 1 \right\},$$
$$\mathbb{S}_{\mathcal{R},n}^M := \left\{ P \in \Pi_{\mathcal{R},n}^M : \ \|P\|_M = 1 \right\}.$$

Further, we define the subsets

$$\mathcal{L}_n^M := \left\{ P \in \mathbb{S}_n^M : \ \varepsilon_p^M(P) > \lambda_1 \right\},$$
$$\mathcal{L}_n^{M,m} := \left\{ P \in \mathbb{S}_{m,n}^M : \ \varepsilon_p^M(P) > \lambda_1 \right\},$$
$$\mathcal{L}_n^{M,\mathcal{R}} := \left\{ P \in \mathbb{S}_{\mathcal{R},n}^M : \ \varepsilon_p^M(P) > \lambda_1 \right\},$$

where $\lambda_1 = \frac{2\beta_M - d + 2}{2\beta_M + d + 2}$ corresponds to the sole root of the Jacobi polynomial $p_1^{\left(\frac{d-2}{2}, \beta_M \right)}(x)$.

Similar as in Section 3.1, we want to use the generalized mean value $\varepsilon_p^M(P)$ and the position variance $\text{var}_{\mathbb{S},p}^M(P)$ as auxiliary tools to determine whether a spherical polynomial

$P \in \mathbb{S}_n^M$ is localized at the point $p \in M$ or not. If $\|P\|_M = 1$, then $\varepsilon_p^M(P)$ is a real value between -1 and 1, and the more the mass of P is concentrated at the point p the closer the value $\varepsilon_p^M(P)$ gets to 1. Hence, we call a spherical polynomial $P \in \mathbb{S}_n^M$ localized at $p \in M$ if $\varepsilon_p^M(P)$ approaches the value 1. Similar as in the case of the Jacobi polynomials, we want to find the following optimally space localized spherical polynomials on M:

$$\mathcal{P}_n^M := \arg \max_{P \in \mathbb{S}_n^M} \varepsilon_p^M(P), \tag{3.61}$$

$$\mathcal{P}_{m,n}^M := \arg \max_{P \in \mathbb{S}_{m,n}^M} \varepsilon_p^M(P), \tag{3.62}$$

$$\mathcal{P}_{\mathcal{R},n}^M := \arg \max_{P \in \mathbb{S}_{\mathcal{R},n}^M} \varepsilon_p^M(P). \tag{3.63}$$

In terms of the position variance $\mathrm{var}_{S,p}^M$ the latter optimization problems can be reformulated as follows:

Proposition 3.24.
Assume that the sets $\mathcal{L}_n^M$, $\mathcal{L}_n^{M,m}$ and $\mathcal{L}_n^{M,\mathcal{R}}$ are nonempty, then

$$\mathcal{P}_n^M = \arg \max_{P \in \mathcal{L}_n^M} \varepsilon_p^M(P) = \arg \min_{P \in \mathcal{L}_n^M} \mathrm{var}_{S,p}^M(P),$$

$$\mathcal{P}_{m,n}^M = \arg \max_{P \in \mathcal{L}_n^{M,m}} \varepsilon_p^M(P) = \arg \min_{P \in \mathcal{L}_n^{M,m}} \mathrm{var}_{S,p}^M(P),$$

$$\mathcal{P}_{\mathcal{R},n}^M = \arg \max_{P \in \mathcal{L}_n^{M,\mathcal{R}}} \varepsilon_p^M(P) = \arg \min_{P \in \mathcal{L}_n^{M,\mathcal{R}}} \mathrm{var}_{S,p}^M(P).$$

Proof. The proof of Proposition 3.24 is identical to the proof of Proposition 3.7 with $\varepsilon_{\alpha\beta}$ and $\mathrm{var}_S^{\alpha\beta}$ replaced by ε_p^M and $\mathrm{var}_{S,p}^M$. $\square$

Hence, by Proposition 3.24, minimizing the position variance $\mathrm{var}_{S,p}^M(P)$ with respect to a polynomial $P \in \mathcal{L}_n^M$ (or $P \in \mathcal{L}_n^{M,m}, \mathcal{L}_n^{M,\mathcal{R}}$, respectively) is equivalent to maximize the mean value $\varepsilon_p^M(P)$. Similar as in Proposition 3.24, the non-emptiness of the sets $\mathcal{L}_n^{M,m}$ and $\mathcal{L}_n^{M,\mathcal{R}}$ can not be guaranteed in general. Since the radial functions on M have an expansion in terms of the Jacobi polynomials $p_n^{(\frac{d-2}{2}, \beta_M)}$ the non-emptiness of the set $\mathcal{L}_n^M$, $n \geq 1$, follows as in Remark 3.8 from the interlacing property of the zeros of the Jacobi polynomials.

To compute the optimally space localized spherical polynomials $\mathcal{P}_n^M$, $\mathcal{P}_{m,n}^M$ and $\mathcal{P}_{\mathcal{R},n}^M$, we need, similarly as in the case of the Jacobi setting (see Lemma 3.5), a characterization of the mean value $\varepsilon_p^M(P)$ in terms of the expansion coefficients $c_{l,k,j}$. Since this characterization will depend on the type of the compact two-point homogeneous space M, we will split the statement into three lemmas.

Lemma 3.25.

Let M be one of the projective spaces $\mathbb{CP}_r^d$, $d \geq 4$, or $\mathbb{HP}_r^d$, $d \geq 8$, or $\mathbb{Ca}_r$, and P a spherical polynomial given by $P^*(t,\xi) = \sum_{l=0}^{n}\sum_{k=0}^{l}\sum_{j=1}^{N(d-1,k)} c_{l,k,j} P_{l,k,j}^{M*}(t,\xi)$. Then,

$$\varepsilon_p^M(P) = \sum_{k=0}^{n}\sum_{j=1}^{N(d-1,k)} \mathbf{c}_{k,j}^H \left[\mathbf{J}\left(\tfrac{d-2}{2}+2k,\beta_M\right)_{n-k}\right]\mathbf{c}_{k,j}, \qquad \text{if} \quad P \in \Pi_n^M,$$

$$\varepsilon_p^M(P) = \sum_{k=0}^{m}\sum_{j=1}^{N(d-1,k)} \mathbf{c}_{k,j}^H \left[\mathbf{J}\left(\tfrac{d-2}{2}+2k,\beta_M\right)_{n-k}^{m-k}\right]\mathbf{c}_{k,j}$$

$$+ \sum_{k=m+1}^{n}\sum_{j=1}^{N(d-1,k)} \mathbf{c}_{k,j}^H \left[\mathbf{J}\left(\tfrac{d-2}{2}+2k,\beta_M\right)_{n-k}\right]\mathbf{c}_{k,j}, \qquad \text{if} \quad P \in \Pi_{m,n}^M,$$

$$\varepsilon_p^M(P) = \tilde{\mathbf{c}}_{0,1}^H \left[\mathbf{J}\left(\tfrac{d-2}{2},\beta_M\right)_n^{m-1}\right]\tilde{\mathbf{c}}_{0,1} + (\varepsilon_p^M(\mathcal{R}) - a_{m-1})|c_{m-1}|^2$$

$$+ \sum_{k=1}^{m}\sum_{j=1}^{N(d-1,k)} \mathbf{c}_{k,j}^H \left[\mathbf{J}\left(\tfrac{d-2}{2}+2k,\beta_M\right)_{n-k}^{m-k}\right]\mathbf{c}_{k,j}$$

$$+ \sum_{k=m+1}^{n}\sum_{j=1}^{N(d-1,k)} \mathbf{c}_{k,j}^H \left[\mathbf{J}\left(\tfrac{d-2}{2}+2k,\beta_M\right)_{n-k}\right]\mathbf{c}_{k,j}, \qquad \text{if} \quad P \in \Pi_{\mathcal{R},n}^M,$$

with the coefficient vectors

$$\mathbf{c}_{k,j} = (c_{m,k,j}, c_{m+1,k,j} \ldots, c_{n,k,j})^T, \quad 0 \leq k \leq m,$$
$$\mathbf{c}_{k,j} = (c_{k,k,j}, c_{k+1,k,j} \ldots, c_{n,k,j})^T, \quad m+1 \leq k \leq n,$$
$$\tilde{\mathbf{c}}_{0,1} = (c_{m-1}, c_{m,0,1}, c_{m+1,0,1} \ldots, c_{n,0,1})^T,$$

and the matrices $\mathbf{J}(\tfrac{d-2}{2}+2k,\beta_M)_n^m$ corresponding to the Jacobi matrices $\mathbf{J}_n^m$ of the associated Jacobi polynomials $p_l^{(\frac{d-2}{2}+2k,\beta_M)}(x,m)$.

Proof. We start with the spherical wavelets $P \in \Pi_{m,n}^M$ having an expansion of the form

$$P^*(t,\xi) = \sum_{l=m}^{n}\sum_{k=0}^{l}\sum_{j=1}^{N(d-1,k)} c_{l,k,j} P_{l,k,j}^{M*}(t,\xi).$$

Taking the definition (3.53) of the spherical polynomials $P_{l,k,j}^M$, we get for the value $\varepsilon_p^M(P)$:

$$\varepsilon_p^M(P) = \int_{\mathbb{S}_p}\int_0^{r\pi} \cos(\tfrac{t}{r})|P^*(t,\xi)|^2 (2r)^{d-1}\sin^{d-1}(\tfrac{t}{2r})\cos^{2\beta_M+1}(\tfrac{t}{2r})\,dt\,d\mu(\xi)$$

$$= \int_{\mathbb{S}_p}\int_0^{r\pi} \left(\sum_{l=m}^{n}\sum_{k=0}^{l}\sum_{j=1}^{N(d-1,k)} c_{l,k,j} 2^k \sin^{2k}(\tfrac{t}{2r})\cos(\tfrac{t}{r})p_{l-k}^{(\frac{d-2}{2}+2k,\beta_M)}(\cos(\tfrac{t}{r}))Y_{k,j}^{d-1}(\xi)\right)$$

$$\times \overline{\left(\sum_{l=m}^{n}\sum_{k=0}^{l}\sum_{j=1}^{N(d-1,k)} c_{l,k,j} 2^k \sin^{2k}(\tfrac{t}{2r})p_{l-k}^{(\frac{d-2}{2}+2k,\beta_M)}(\cos(\tfrac{t}{r}))Y_{k,j}^{d-1}(\xi)\right)}$$

$$\times 2^{\frac{d}{2}+\beta_M}\tfrac{1}{r}\sin^{d-1}(\tfrac{t}{2r})\cos^{2\beta_M+1}(\tfrac{t}{2r})\,dt\,d\mu(\xi).$$

Now, we use the orthonormality of the spherical harmonics $Y_{k,j}^{d-1}$ and rearrange the order of summation. Then, we get

$$
\varepsilon_p^M(P) = \sum_{k=0}^{m} \sum_{j=1}^{N(d-1,k)} \int_0^{r\pi} \left(\sum_{l=m}^{n} c_{l,k,j} \cos(\tfrac{t}{r}) p_{l-k}^{(\frac{d-2}{2}+2k,\beta_M)}(\cos(\tfrac{t}{r})) \right)
$$
$$
\times \overline{\left(\sum_{l=m}^{n} c_{l,k,j} p_{l-k}^{(\frac{d-2}{2}+2k,\beta_M)}(\cos(\tfrac{t}{r})) \right)} 2^{2k+\frac{d}{2}+\beta_M} \tfrac{1}{r} \sin^{4k+d-1}(\tfrac{t}{2r}) \cos^{2\beta_M+1}(\tfrac{t}{2r}) dt
$$
$$
+ \sum_{k=m+1}^{n} \sum_{j=1}^{N(d-1,k)} \int_0^{r\pi} \left(\sum_{l=k}^{n} c_{l,k,j} \cos(\tfrac{t}{r}) p_{l-k}^{(\frac{d-2}{2}+2k,\beta_M)}(\cos(\tfrac{t}{r})) \right)
$$
$$
\times \overline{\left(\sum_{l=k}^{n} c_{l,k,j} p_{l-k}^{(\frac{d-2}{2}+2k,\beta_M)}(\cos(\tfrac{t}{r})) \right)} 2^{2k+\frac{d}{2}+\beta_M} \tfrac{1}{r} \sin^{4k+d-1}(\tfrac{t}{2r}) \cos^{2\beta_M+1}(\tfrac{t}{2r}) dt.
$$

Finally, using the three-term recurrence relation (3.9) and the orthonormality of the Jacobi polynomials $p_l^{(\frac{d-2}{2}+2k,\beta_M)}$, we conclude

$$
\varepsilon_p^M(P) = \sum_{k=0}^{m} \sum_{j=1}^{N(d-1,k)} \int_0^{r\pi} \left(\sum_{l=m}^{n} c_{l,k,j} \left(b_{l-k} p_{l-k-1}^{(\frac{d-2}{2}+2k,\beta_M)}(\cos(\tfrac{t}{r})) + a_l p_{l-k}^{(\frac{d-2}{2}+2k,\beta_M)}(\cos(\tfrac{t}{r})) \right) \right.
$$
$$
\left. + b_{l-k+1} p_{l-k+1}^{(\frac{d-2}{2}+2k,\beta_M)}(\cos(\tfrac{t}{r})) \right) \overline{\left(\sum_{l=m}^{n} c_{l,k,j} p_{l-k}^{(\frac{d-2}{2}+2k,\beta_M)}(\cos(\tfrac{t}{r})) \right)}
$$
$$
\times 2^{2k+\frac{d}{2}+\beta_M} \tfrac{1}{r} \sin^{4k+d-1}(\tfrac{t}{2r}) \cos^{2\beta_M+1}(\tfrac{t}{2r}) dt
$$
$$
+ \sum_{k=m+1}^{n} \sum_{j=1}^{N(d-1,k)} \int_0^{r\pi} \left(\sum_{l=k}^{n} c_{l,k,j} \left(b_{l-k} p_{l-k-1}^{(\frac{d-2}{2}+2k,\beta_M)}(\cos(\tfrac{t}{r})) + a_l p_{l-k}^{(\frac{d-2}{2}+2k,\beta_M)}(\cos(\tfrac{t}{r})) \right) \right.
$$
$$
\left. + b_{l-k+1} p_{l-k+1}^{(\frac{d-2}{2}+2k,\beta_M)}(\cos(\tfrac{t}{r})) \right) \overline{\left(\sum_{l=k}^{n} c_{l,k,j} p_{l-k}^{(\frac{d-2}{2}+2k,\beta_M)}(\cos(\tfrac{t}{r})) \right)}
$$
$$
\times 2^{2k+\frac{d}{2}+\beta_M} \tfrac{1}{r} \sin(\tfrac{t}{2r})^{4k+d-1} \cos(\tfrac{t}{2r})^{2\beta_M+1} dt
$$
$$
= \sum_{k=0}^{m} \sum_{j=1}^{N(d-1,k)} \mathbf{c}_{k,j}^H \left[\mathbf{J}\left(\tfrac{d-2}{2}+2k, \beta_M\right)_{n-k}^{m-k} \right] \mathbf{c}_{k,j}
$$
$$
+ \sum_{k=m+1}^{n} \sum_{j=1}^{N(d-1,k)} \mathbf{c}_{k,j}^H \left[\mathbf{J}\left(\tfrac{d-2}{2}+2k, \beta_M\right)_{n-k} \right] \mathbf{c}_{k,j}.
$$

Thus, the statement for the spherical wavelets $P \in \Pi_{m,n}^M$ is shown. If $m = 0$, the statement for $P \in \Pi_n^M$ follows as a special case. Finally, the statement for $P \in \Pi_{\mathcal{R},n}^M$ follows, if we set $m = 0$ and consider coefficient sets $\{c_{l,k,j}\}$ of the form

$$
\begin{aligned}
c_{l,k,j} &\in \mathbb{C}, && \text{if } l \geq m, \\
c_{l,k,j} &= 0, && \text{if } l < m,\ k \neq 0, \\
c_{l,k,j} &= e_l c_{m-1}, && \text{if } l < m,\ k = 0,
\end{aligned}
$$

where e_l denote the expansion coefficients of the radial function $\mathcal{R}$. $\qquad\square$

3. Optimally space localized polynomials

Lemma 3.26.
Let $M = \mathbb{S}_r^d$ and P be given by $P^*(t,\xi) = \sum_{l=0}^n \sum_{k=0}^l \sum_{j=1}^{N(d-1,k)} c_{l,k,j} P_{l,k,j}^{\mathbb{S}_r^{d*}}(t,\xi)$. Then,

$$\varepsilon_p^{\mathbb{S}_r^d}(P) = \sum_{k=0}^n \sum_{j=1}^{N(d-1,k)} \mathbf{c}_{k,j}^H \left[\mathbf{J}\left(\tfrac{d-2}{2}+k, \tfrac{d-2}{2}+k\right)_{n-k} \right] \mathbf{c}_{k,j}, \qquad \text{if} \quad P \in \Pi_n^{\mathbb{S}_r^d},$$

$$\varepsilon_p^{\mathbb{S}_r^d}(P) = \sum_{k=0}^m \sum_{j=1}^{N(d-1,k)} \mathbf{c}_{k,j}^H \left[\mathbf{J}\left(\tfrac{d-2}{2}+k, \tfrac{d-2}{2}+k\right)_{n-k}^{m-k} \right] \mathbf{c}_{k,j}$$

$$+ \sum_{k=m+1}^n \sum_{j=1}^{N(d-1,k)} \mathbf{c}_{k,j}^H \left[\mathbf{J}\left(\tfrac{d-2}{2}+k, \tfrac{d-2}{2}+k\right)_{n-k} \right] \mathbf{c}_{k,j}, \qquad \text{if} \quad P \in \Pi_{m,n}^{\mathbb{S}_r^d},$$

$$\varepsilon_p^{\mathbb{S}_r^d}(P) = \tilde{\mathbf{c}}_{0,1}^H \left[\mathbf{J}\left(\tfrac{d-2}{2}, \tfrac{d-2}{2}\right)_n^{m-1} \right] \tilde{\mathbf{c}}_{0,1} + (\varepsilon_p^{\mathbb{S}_r^d}(\mathcal{R}) - a_{m-1})|c_{m-1}|^2$$

$$+ \sum_{k=1}^m \sum_{j=1}^{N(d-1,k)} \mathbf{c}_{k,j}^H \left[\mathbf{J}\left(\tfrac{d-2}{2}+k, \tfrac{d-2}{2}+k\right)_{n-k}^{m-k} \right] \mathbf{c}_{k,j}$$

$$+ \sum_{k=m+1}^n \sum_{j=1}^{N(d-1,k)} \mathbf{c}_{k,j}^H \left[\mathbf{J}\left(\tfrac{d-2}{2}+k, \tfrac{d-2}{2}+k\right)_{n-k} \right] \mathbf{c}_{k,j}, \qquad \text{if} \quad P \in \Pi_{\mathcal{R},n}^{\mathbb{S}_r^d},$$

with the coefficient vectors

$$\mathbf{c}_{k,j} = (c_{m,k,j}, c_{m+1,k,j} \ldots, c_{n,k,j})^T, \quad 0 \le k \le m,$$
$$\mathbf{c}_{k,j} = (c_{k,k,j}, c_{k+1,k,j} \ldots, c_{n,k,j})^T, \quad m+1 \le k \le n,$$
$$\tilde{\mathbf{c}}_{0,1} = (c_{m-1}, c_{m,0,1}, c_{m+1,0,1} \ldots, c_{n,0,1})^T,$$

and the matrices $\mathbf{J}(\tfrac{d-2}{2}+k, \tfrac{d-2}{2}+k)_n^m$ corresponding to the Jacobi matrices $\mathbf{J}_n^m$ of the associated ultraspherical polynomials $p_l^{(\frac{d-2}{2}+k,\frac{d-2}{2}+k)}(x,m)$.

Proof. Up to an adaption of the underlying basis, the proof of Lemma 3.26 is the same as the proof of Lemma 3.25. The details are therefore omitted. $\square$

Lemma 3.27.
If $M = \mathbb{RP}_r^d$ is the real projective space and P is a spherical polynomial on $\mathbb{RP}_r^d$ defined by $P^*(t,\xi) = \sum_{l=0}^n \sum_{k=0}^l \sum_{j=1}^{N(d-1,k)} c_{l,k,j} P_{l,k,j}^{\mathbb{RP}_r^{d*}}(t,\xi)$, we get

$$\varepsilon_p^{\mathbb{RP}_r^d}(P) = \sum_{k=0}^n \sum_{j=1}^{N(d-1,k)} \mathbf{c}_{k,j}^H \left[\mathbf{J}\left(\tfrac{d-2}{2}+k, -\tfrac{(-1)^k}{2}\right)_{n-k} \right] \mathbf{c}_{k,j}, \qquad \text{if} \quad P \in \Pi_n^{\mathbb{RP}_r^d},$$

$$\varepsilon_p^{\mathbb{RP}_r^d}(P) = \sum_{k=0}^m \sum_{j=1}^{N(d-1,k)} \mathbf{c}_{k,j}^H \left[\mathbf{J}\left(\tfrac{d-2}{2}+k, -\tfrac{(-1)^k}{2}\right)_{n-k}^{m-k} \right] \mathbf{c}_{k,j}$$

$$+ \sum_{k=m+1}^n \sum_{j=1}^{N(d-1,k)} \mathbf{c}_{k,j}^H \left[\mathbf{J}\left(\tfrac{d-2}{2}+k, -\tfrac{(-1)^k}{2}\right)_{n-k} \right] \mathbf{c}_{k,j}, \qquad \text{if} \quad P \in \Pi_{m,n}^{\mathbb{RP}_r^d},$$

$$\varepsilon_p^{\mathbb{RP}_r^d}(P) = \tilde{\mathbf{c}}_{0,1}^H \left[\mathbf{J}\left(\tfrac{d-2}{2}, -\tfrac{1}{2}\right)_n^{m-1} \right] \tilde{\mathbf{c}}_{0,1} + \gamma_{m-1} |c_{m-1}|^2$$

$$+ \sum_{k=1}^{m} \sum_{j=1}^{N(d-1,k)} \mathbf{c}_{k,j}^H \left[\mathbf{J}\left(\tfrac{d-2}{2} + k, -\tfrac{(-1)^k}{2}\right)_{n-k}^{m-k} \right] \mathbf{c}_{k,j}$$

$$+ \sum_{k=m+1}^{n} \sum_{j=1}^{N(d-1,k)} \mathbf{c}_{k,j}^H \left[\mathbf{J}\left(\tfrac{d-2}{2} + k, -\tfrac{(-1)^k}{2}\right)_{n-k} \right] \mathbf{c}_{k,j}, \qquad if \quad P \in \Pi_{\mathcal{R},n}^{\mathbb{RP}_r^d},$$

with the coefficient vectors $\tilde{\mathbf{c}}_{0,1}$ and $\mathbf{c}_{k,j}$ as defined in Lemma 3.26 and the matrices $\mathbf{J}(\tfrac{d-2}{2} + k, -\tfrac{(-1)^k}{2})_n^m$ corresponding to the Jacobi matrices $\mathbf{J}_n^m$ of the associated Jacobi polynomials $p_l^{(\frac{d-2}{2}+k, -\frac{(-1)^k}{2})}(x,m)$.

Proof. Using the orthonormal spherical polynomials (3.52) of the real projective space $\mathbb{RP}_r^d$ instead of the orthonormal spherical polynomials in Lemma 3.25, the proof of Lemma 3.27 is up to a minor change in the notion the same as the proof of Lemma 3.25. So, the details are omitted also at this place. □

Now, for the optimization problems (3.61), (3.62) and (3.63), we can conclude:

Theorem 3.28.
Let $P \in \mathbb{S}_n^M, \mathbb{S}_{m,n}^M, \mathbb{S}_{\mathcal{R},n}^M$, respectively. Then, the maximum of $\varepsilon_p^M(P)$ is attained for the radial spherical polynomials

$$\mathcal{P}_n^{M*}(t) = \kappa_1 \sum_{l=0}^{n} p_l^{(\alpha,\beta)}(\lambda_{n+1}) P_l^M(t), \tag{3.64}$$

$$\mathcal{P}_n^{M,m*}(t) = \kappa_2 \sum_{l=m}^{n} p_{l-m}^{(\alpha,\beta)}(\lambda_{n-m+1}^m, m) P_l^M(t), \tag{3.65}$$

$$\mathcal{P}_n^{M,\mathcal{R}*}(t) = \kappa_3 \left(\mathcal{R}(t) + \sum_{l=m}^{n} p_{l-m+1}^{(\alpha,\beta)}(\lambda_{n-m+2}^{\mathcal{R}}, m-1, \gamma_{\mathcal{R}}, \delta_{\mathcal{R}}) P_l^M(t) \right), \tag{3.66}$$

where the values λ_{n+1}, λ_{n-m+1}^m and $\lambda_{n-m+2}^{\mathcal{R}}$ denote the largest zero of the polynomials $p_{n+1}^{(\frac{d-2}{2}, \beta_M)}(x)$, $p_{n-m+1}^{(\frac{d-2}{2}, \beta_M)}(x,m)$ and $p_{n-m+2}^{(\frac{d-2}{2}, \beta_M)}(x, m-1, \gamma_{\mathcal{R}}, \delta_{\mathcal{R}})$ in $[-1,1]$, respectively. The parameters of the scaled co-recursive associated polynomials $p_{l-m+1}^{(\alpha,\beta)}(x, m-1, \gamma_{\mathcal{R}}, \delta_{\mathcal{R}})$ are given as $\gamma_{\mathcal{R}} = \varepsilon_p^M(\mathcal{R}) - a_{m-1}$ and $\delta_{\mathcal{R}} = \|\mathcal{R}\|_M^2$. The constants κ_1, κ_2 and κ_3 are normalization factors which ensure that $\|\mathcal{P}_n^M\|_M = \|\mathcal{P}_{m,n}^M\|_M = \|\mathcal{P}_{\mathcal{R},n}^M\|_M = 1$ is satisfied. The constants κ_1, κ_2 and κ_3 are uniquely determined up to multiplication with a complex scalar of absolute value one. Finally, the maximum values of $\varepsilon_{\alpha\beta}(P)$ are given as

$$M_n^M := \max_{P \in \mathbb{S}_n^M} \varepsilon_p^M(P) = \lambda_{n+1},$$

$$M_{m,n}^M := \max_{P \in \mathbb{S}_{m,n}^M} \varepsilon_p^M(P) = \lambda_{n-m+1}^m,$$

$$M_{\mathcal{R},n}^M := \max_{P \in \mathbb{S}_{\mathcal{R},n}^M} \varepsilon_p^M(P) = \lambda_{n-m+2}^{\mathcal{R}}.$$

Proof. We consider first the case when M is one the projective spaces $\mathbb{CP}_r^d$, $d \geq 4$, or $\mathbb{HP}_r^d$, $d \geq 8$, or $\mathbb{C}a_r$, and $P \in \mathbb{S}_{m,n}^M$ is a spherical wavelet given by $P^*(t, \xi) = \sum_{l=m}^n \sum_{k=0}^l \sum_{j=1}^{N(d-1,k)} c_{l,k,j} P_{l,k,j}^{M*}(t, \xi)$. Then, by Lemma 3.25, we can write the mean value $\varepsilon_p^M(P)$ as

$$\varepsilon_p^M(P) = \sum_{k=0}^m \sum_{j=1}^{N(d-1,k)} \mathbf{c}_{k,j}^H \left[\mathbf{J}\left(\tfrac{d-2}{2} + 2k, \beta_M\right)_{n-k}^{m-k} \right] \mathbf{c}_{k,j}$$
$$+ \sum_{k=m+1}^n \sum_{j=1}^{N(d-1,k)} \mathbf{c}_{k,j}^H \left[\mathbf{J}\left(\tfrac{d-2}{2} + 2k, \beta_M\right)_{n-k} \right] \mathbf{c}_{k,j}, \qquad (3.67)$$

with the coefficient vectors

$$\mathbf{c}_{k,j} = (c_{m,k,j}, c_{m+1,k,j}, \dots, c_{n,k,j})^T, \quad 0 \leq k \leq m,$$
$$\mathbf{c}_{k,j} = (c_{k,k,j}, c_{k+1,k,j}, \dots, c_{n,k,j})^T, \quad m+1 \leq k \leq n,$$

and the matrices $\mathbf{J}(\tfrac{d-2}{2} + 2k, \beta_M)_{n-k}^{m-k}$ corresponding to the Jacobi matrices of the associated Jacobi polynomials $p_l^{(\tfrac{d-2}{2}+2k,\beta_M)}(x, m-k)$.

Now, maximizing $\varepsilon_p^M(P)$ with respect to $P \in \mathbb{S}_{m,n}^M$ is equivalent to maximize the quadratic functional (3.67) subject to $\sum_{k=0}^n \sum_{j=1}^{N(d-1,k)} |\mathbf{c}_{k,j}|^2 = 1$. Hence, we get

$$\sum_{k=0}^m \sum_{j=1}^{N(d-1,k)} \mathbf{c}_{k,j}^H \left[\mathbf{J}\left(\tfrac{d-2}{2} + k, \tfrac{d-2}{2} + k\right)_{n-k}^{m-k} \right] \mathbf{c}_{k,j} \qquad (3.68)$$
$$+ \sum_{k=m+1}^n \sum_{j=1}^{N(d-1,k)} \mathbf{c}_{k,j}^H \left[\mathbf{J}\left(\tfrac{d-2}{2} + k, \tfrac{d-2}{2} + k\right)_{n-k} \right] \mathbf{c}_{k,j} \leq \lambda_{\max} \sum_{k=0}^n \sum_{j=1}^{N(d-1,k)} |\mathbf{c}_{k,j}|_2^2,$$

where $\lambda_{\max}$ corresponds to the largest eigenvalue taken over all the symmetric matrices $\mathbf{J}\left(\tfrac{d-2}{2} + 2k, \beta_M\right)_{n-k}$ and $\mathbf{J}\left(\tfrac{d-2}{2} + 2k, \beta_M\right)_{n-k}^{m-k}$ in (3.68), and where equality holds for an eigenvector corresponding to $\lambda_{\max}$. Moreover, the eigenvalues of the matrices $\mathbf{J}(\tfrac{d-2}{2} + 2k, \beta_M)_{n-k}$ correspond to the roots of the Jacobi polynomials $p_{n-k+1}^{(\tfrac{d-2}{2}+2k,\beta_M)}(x)$ and the eigenvalues of the matrices $\mathbf{J}(\tfrac{d-2}{2} + 2k, \beta_M)_{n-k}^{m-k}$ to the zeros of the associated polynomials $p_{n-m+1}^{(\tfrac{d-2}{2}+2k,\beta_M)}(x, m-k)$.

Now, the results of Section 3.2 on the monotonicity of the largest zero of associated Jacobi polynomials come into play. Due to the interlacing property of the zeros of the Jacobi polynomials (see [83, Theorem 3.3.2]) and Corollary 3.18 (alternatively, one could also use the results in [13] and [44]), the matrix $\mathbf{J}(\tfrac{d-2}{2} + 2k, \beta_M)_{n-k}$, $m \leq k \leq n$, with the largest eigenvalue is precisely the matrix $\mathbf{J}(\tfrac{d-2}{2} + 2m, \beta_M)_{n-m}$. Further, by Corollary 3.20, the matrix $\mathbf{J}(\tfrac{d-2}{2} + 2k, \beta_M)_{n-k}^{m-k}$, $0 \leq k \leq m$, with the largest eigenvalue is the matrix $\mathbf{J}(\tfrac{d-2}{2}, \beta_M)_n^m$ which appears only one time as a submatrix in (3.67). Hence, the unique overall submatrix in (3.67) with the largest eigenvalue is precisely the matrix $\mathbf{J}(\tfrac{d-2}{2}, \beta_M)_n^m$ and $\lambda_{\max} = \lambda_{n+m+1}^m$ corresponds to the largest zero of the associated

Jacobi polynomial $p_{n-m+1}^{(\frac{d-2}{2},\beta_M)}(x,m)$. Due to the three-term recurrence relation (3.12) of the polynomial $p_{n-m+1}^{(\frac{d-2}{2},\beta_M)}(x,m)$, the coefficients of the corresponding eigenvector can be determined as

$$c_{l,0,1} = p_{l-m}^{(\frac{d-2}{2},\beta_M)}(\lambda_{n-m+1}^m,m), \quad \text{if} \quad m \le l \le n,$$
$$c_{l,k,j} = 0, \quad \text{if} \quad 0 \le l \le m-1 \quad \text{or} \quad k \ne 0.$$

Next, we have to normalize the coefficients $c_{l,k,j}$ such that $\sum_{l=m}^n |p_{l-m}^{(\frac{d-2}{2},\beta_M)}(\lambda_{n-m+1}^m,m)|^2 = 1$. This is done by an appropriately defined constant κ_2. The uniqueness (up to a complex scalar with absolute value 1) of the optimal polynomial $\mathcal{P}_{m,n}^M$ follows from the fact that the largest zero of $p_{n-m+1}^{(\frac{d-2}{2},\beta_M)}(x,m)$ is simple and that the largest eigenvalue of the matrix $\mathbf{J}(\frac{d-2}{2},\beta_M)_n^m$ is strictly larger than the largest eigenvalues of all other submatrices in (3.67) (see Corollary 3.20). The formula for $M_{m,n}^M$ follows directly from the estimate (3.68). Moreover, if we set $m = 0$, we get the formula for $\mathcal{P}_n^M$ as a special case.

Next, we will show the formula for $\mathcal{P}_{\mathcal{R},n}^M$. Lemma 3.25 states that the mean value $\varepsilon_p^M(P)$ of a polynomial $P \in \Pi_{\mathcal{R},n}^M$ given by $P^*(t,\xi) = c_{m-1}\mathcal{R}(t) + \sum_{l=m}^n \sum_{k=0}^l \sum_{j=1}^{N(d-1,k)} c_{l,k,j} P_{l,k,j}^{M*}(t,\xi)$ can be written as

$$\varepsilon_p^M(P) = \tilde{\mathbf{c}}_{0,1}^H \left[\mathbf{J}\left(\frac{d-2}{2},\beta_M\right)_n^{m-1}\right]\tilde{\mathbf{c}}_{0,1} + (\varepsilon_p^M(\mathcal{R}) - a_{m-1})|c_{m-1}|^2 \qquad (3.69)$$
$$+ \sum_{k=1}^m \sum_{j=1}^{N(d-1,k)} \mathbf{c}_{k,j}^H \left[\mathbf{J}\left(\frac{d-2}{2}+2k,\beta_M\right)_{n-k}^{m-k}\right]\mathbf{c}_{k,j}$$
$$+ \sum_{k=m+1}^n \sum_{j=1}^{N(d-1,k)} \mathbf{c}_{k,j}^H \left[\mathbf{J}\left(\frac{d-2}{2}+2k,\beta_M\right)_{n-k}\right]\mathbf{c}_{k,j},$$

with the coefficient vectors

$$\mathbf{c}_{k,j} = (c_{m,k,j}, c_{m+1,k,j}, \dots, c_{n,k,j})^T, \quad 1 \le k \le m,$$
$$\mathbf{c}_{k,j} = (c_{k,k,j}, c_{k+1,k,j}, \dots, c_{n,k,j})^T, \quad m+1 \le k \le n,$$
$$\tilde{\mathbf{c}}_{0,1} = (c_{m-1}, c_{m,0,1}, c_{m+1,0,1}, \cdots, c_{n,0,1})^T.$$

Maximizing $\varepsilon_p^M(P)$ with respect to a polynomial $P \in \mathbb{S}_{\mathcal{R},n}^M$ is therefore equivalent to maximize the quadratic functional in (3.69) subject to $(\|\mathcal{R}\|_M^2 - 1)|c_{m-1}|^2 + |\tilde{\mathbf{c}}_{0,1}|^2 + \sum_{k=1}^n \sum_{j=1}^{N(d-1,k)} |\mathbf{c}_{k,j}|^2 = 1$. Using a Lagrange multiplier λ and differentiating the Lagrange function with respect to the coefficients $c_{l,k,j}$, we obtain a block matrix equation with the linear subsystems of equations

$$\left[\mathbf{J}\left(\frac{d-2}{2},\beta_M\right)_n^{m-1}\right]\tilde{\mathbf{c}}_{0,1} + \gamma_{\mathcal{R}}(c_{m-1},0,\dots,0)^T = \lambda\left(\delta_{\mathcal{R}}c_{m-1}, c_{m,0,1}, c_{m+1,0,1}, \dots, c_{n,0,1}\right)^T,$$
$$\left[\mathbf{J}\left(\frac{d-2}{2}+2k,\beta_M\right)_{n-k}^{m-k}\right]\mathbf{c}_{k,j} = \lambda\mathbf{c}_{k,j}, \quad 1 \le k \le m, \ 1 \le j \le N(d-1,k), \qquad (3.70)$$
$$\left[\mathbf{J}\left(\frac{d-2}{2}+2k,\beta_M\right)_{n-k}\right]\mathbf{c}_{k,j} = \lambda\mathbf{c}_{k,j}, \quad m+1 \le k \le n, \ 1 \le j \le N(d-1,k),$$

as necessary conditions for the maximum, where $\gamma_{\mathcal{R}} = \varepsilon_p^M(\mathcal{R}) - a_{m-1}$ and $\delta_{\mathcal{R}} = \|\mathcal{R}\|_M^2$. In (3.70), the first system of equations is equivalent to the symmetric eigenvalue problem

$$\tilde{\mathbf{J}}(\tfrac{d-2}{2}, \beta_M)_n^{m-1} \tilde{\tilde{\mathbf{c}}}_{0,1} = \lambda \tilde{\tilde{\mathbf{c}}}_{0,1}, \tag{3.71}$$

with the symmetric matrix

$$\tilde{\mathbf{J}}(\tfrac{d-2}{2}, \beta_M)_n^{m-1} = \begin{pmatrix} \frac{a_{m-1}+\gamma_{\mathcal{R}}}{\sqrt{\delta_{\mathcal{R}}}} & \frac{b_m}{\sqrt{\delta_{\mathcal{R}}}} & 0 & \cdots & 0 \\ \frac{b_m}{\sqrt{\delta_{\mathcal{R}}}} & & & & \\ 0 & & \mathbf{J}(\tfrac{d-2}{2}, \beta_M)_n^m & & \\ \vdots & & & & \\ 0 & & & & \end{pmatrix}$$

and the coefficient vector

$$\tilde{\tilde{\mathbf{c}}}_{0,1} = \left(\sqrt{\delta_{\mathcal{R}}} c_{m-1}, c_{m,0,1}, c_{m+1,0,1}, \ldots, c_{n,0,1} \right)^T.$$

From the argumentation after equation (3.67) we know that from the matrices $\mathbf{J}(\tfrac{d-2}{2} + 2k, \beta_M)_{n-k}^{m-k}$, $k = 0, \ldots, m$, and $\mathbf{J}(\tfrac{d-2}{2} + 2k, \beta_M)_{n-k}$, $k = m+1, \ldots, n$, the matrix with the largest eigenvalue is $\mathbf{J}(\tfrac{d-2}{2}, \beta_M)_n^m$. Moreover, because of the eigenvalue interlacing theorem for bordered matrices (see [39, Theorem 4.3.8]), the largest eigenvalue of the matrix $\tilde{\mathbf{J}}(\tfrac{d-2}{2}, \beta_M)_n^{m-1}$ is strictly larger than the largest eigenvalue of $\mathbf{J}(\tfrac{d-2}{2}, \beta_M)_n^m$. Since the eigenvalue equation (3.71) is equivalent to the first system of equations in (3.70), the eigenvalues of $\tilde{\mathbf{J}}(\tfrac{d-2}{2}, \beta_M)_n^{m-1}$ correspond to the zeros of the co-recursive associated Jacobi polynomials $p_{n-m+2}^{(\frac{d-2}{2}, \beta_M)}(x, m-1, \gamma_{\mathcal{R}}, \delta_{\mathcal{R}})$ (cf. the three-term recursion formula (3.13)).

In total, the largest eigenvalue of all matrices in (3.70) corresponds precisely to the largest zero $\lambda_{n-m+2}^{\mathcal{R}}$ of the polynomial $p_{n-m+2}^{(\frac{d-2}{2}, \beta_M)}(x, m-1, \gamma_{\mathcal{R}}, \delta_{\mathcal{R}})$. The corresponding eigenvector can be computed from the three-term recurrence relation (3.13) as

$$c_{m-1} = 1,$$
$$c_{l,0,1} = p_{l-m+1}^{(\frac{d-2}{2}, \beta_M)}(\lambda_{n-m+2}^{\mathcal{R}}, m-1, \gamma_{\mathcal{R}}, \delta_{\mathcal{R}}), \quad \text{for} \quad m \le l \le n,$$
$$c_{l,k,j} = 0, \quad \text{if} \quad m \le l \le n, \quad k \ne 0.$$

The coefficients are normalized by the constant κ_3 and the uniqueness (up to multiplication with a complex scalar of absolute value 1) follows from the fact that the largest eigenvalue of the orthogonal polynomials $p_{n-m+2}^{(\frac{d-2}{2}, \beta_M)}(x, m-1, \gamma_{\mathcal{R}}, \delta_{\mathcal{R}})$ is simple and the fact that the largest eigenvalue of the matrix $\tilde{\mathbf{J}}(\tfrac{d-2}{2}, \beta_M)_n^{m-1}$ is strictly larger than the largest eigenvalues of all the submatrices in (3.70) (see again Corollary 3.20). Finally, the value for $M_{\mathcal{R},n}^M$ follows also from (3.70) and the formula (3.69) for ε_p^M.

If M is the sphere $\mathbb{S}_r^d$ or the real projective space $\mathbb{RP}_r^d$, the proof of Theorem 3.28 is almost identical to the preceding proof. The only difference lies in the fact that in the formulas for the basis polynomials $P_{l,k,j}^M$ (see (3.51) and (3.52)) and in the formulas for the mean value $\varepsilon_p^M(P)$ (see Lemma 3.26 and 3.27) one uses the ultraspherical polynomials for the sphere and particular Jacobi polynomials for the real projective spaces instead of the Jacobi polynomials in the above argumentation. Since otherwise the proof remains conceptually the same, the details are omitted at this place. For $M = \mathbb{RP}^2$ one has to use Theorem 3.21 in addition to Corollary 3.20. □

Since the polynomials $\mathcal{P}_n^M$, $\mathcal{P}_{m,n}^M$ and $\mathcal{P}_{\mathcal{R},n}^M$ are radial functions on M and have an expansion in terms of the Jacobi polynomials $p_l^{(\frac{d-2}{2},\beta_M)}$, we can use Corollary 3.10 to get explicit formulas for the optimally space localized polynomials on M.

Corollary 3.29.
The optimally space localized spherical polynomials $\mathcal{P}_n^M$, $\mathcal{P}_{m,n}^M$ and $\mathcal{P}_{\mathcal{R},n}^M$ of Theorem 3.28 have the explicit representation

$$\mathcal{P}_n^{M*}(t) = \kappa_1 b_{n+1} \frac{P_{n+1}^{M*}(t) p_n^{(\frac{d-2}{2},\beta_M)}(\lambda_{n+1})}{\cos(\frac{t}{r}) - \lambda_{n+1}},$$

$$\mathcal{P}_n^{M,m*}(t) = \kappa_2 \frac{b_{n+1} P_{n+1}^{M*}(t) p_{n-m}^{(\frac{d-2}{2},\beta_M)}(\lambda_{n-m+1}^m, m) + b_m P_{m-1}^{M*}(t)}{\cos(\frac{t}{r}) - \lambda_{n-m+1}^m},$$

$$\mathcal{P}_n^{M,\mathcal{R}*}(t) = \kappa_3 \left(\frac{b_{n+1} P_{n+1}^{M*}(t) p_{n-m+1}^{(\frac{d-2}{2},\beta_M)}(\lambda_{n-m+2}^{\mathcal{R}}, m-1, \gamma_{\mathcal{R}}, \delta_{\mathcal{R}})}{\cos(\frac{t}{r}) - \lambda_{n-m+2}^{\mathcal{R}}} \right.$$
$$\left. + \frac{P_{m-1}^{M*}(t)((\delta_{\mathcal{R}} - 1)\lambda_{n-m+2}^{\mathcal{R}} - \gamma_{\mathcal{R}}) + b_{m-1} P_{m-2}^{M*}(\cos(\frac{t}{r}))}{\cos(\frac{t}{r}) - \lambda_{n-m+2}^{\mathcal{R}}} \right),$$

where the constants κ_1, κ_2, κ_3 and the roots λ_{n+1}, λ_{n-m+1}^m, $\lambda_{n-m+2}^{\mathcal{R}}$ are given as in Theorem 3.28.

Remark 3.30. The Christoffel-Darboux kernel and the de La Vallée Poussin kernel introduced in Section 3.1.3 and Section 3.1.4 for the Jacobi polynomials can be considered also as radial spherical polynomials on the compact two-point homogeneous spaces. For this, we define the kernels K_n^m and V_n^M on M by

$$K_n^{M*}(t) := \frac{1}{|\mathfrak{S}_p|^{\frac{1}{2}} 2^{\frac{d-2-2\beta_M}{4}} r^{\frac{d}{2}}} K_n^{(\frac{d-2}{2},\beta_M)}(\tfrac{t}{r}),$$

$$V_n^{M*}(t) := V_n(\tfrac{t}{r}).$$

In particular, the Christoffel-Darboux kernel K_n^M plays an important role in the theory of polynomial approximation on M. One of its remarkable properties is the so called reproducing property for spherical polynomials $P \in \Pi_n^M$, i.e., the kernel K_n^M satisfies (see [57])

$$P(q) = \int_M P(s) K_n^{M*}(d(s,q)) d\mu_M(s), \quad q \in M, \ P \in \Pi_n^M.$$

Moreover, the operator S_n on $L^2(M)$ given by $S_n f(q) = \int_M f(s) K_n^{M*}(d(s,q)) d\mu_M(s)$ is the orthogonal projection of the function $f \in L^2(M)$ onto the subspace Π_n^M (see [53]).

3.4. Remarks and References

Associated and scaled co-recursive associated polynomials. Associated polynomials $P_l(x, c)$ can be defined quite generally for orthogonal polynomials $P_l(x)$ by shifting the coefficients in the three-term recurrence relation, similar as in (3.12) for the Jacobi polynomials.

For $c = 1$, the associated polynomials $P_l(x, 1)$ are sometimes also called numerator polynomials (see [10, Definition 4.1]). Particular results for some families of associated polynomials like the associated Laguerre, Hermite and Jacobi polynomials concerning orthogonality measures, explicit forms and differential equations can be found in [42, Sections 5.6, 5.7, 15.9], [87] and in further references therein.

Co-recursive associated polynomials were introduced for special cases by J. Letessier in [51], [52]. The polynomials $p_l^{(\alpha,\beta)}(x, m, \gamma, \delta)$ coincide with a slightly different notation with the scaled co-recursive associated Jacobi polynomials considered in [40].

Optimally space localized Jacobi polynomials. Optimally space localized trigonometric polynomials and wavelets that minimize the angular variance of the Breitenberger uncertainty principle (1.18), and, in particular, Example 3.11, were firstly considered by Rauhut in [70] and [71]. In [71], also the limit $n \to \infty$ for the uncertainty product of the optimally space localized trigonometric polynomials (corresponding to the statement of Theorem 3.14 with $\alpha = \beta = -\frac{1}{2}$) was computed.

For the more general Jacobi case, Theorem 3.6 is a novel result. In particular, the polynomial spaces $\Pi_{\mathcal{R},n}^{(\alpha,\beta)}$ that play an important role in the theory of polynomial approximation and the respective optimally space localized polynomials $\mathcal{P}_{\mathcal{R},n}^{(\alpha,\beta)}$ are considered for the first time. New are also the Christoffel-Darboux-type formulas in Lemma 3.9 and the explicit formulas for the optimally space localized polynomials in Corollary 3.10.

Beside the theory discussed in Section 3.1, there exist also other concepts of localization of polynomials in the literature. In particular, in [17], Filbir, Mhaskar and Prestin constructed exponentially localized polynomial kernels ϕ_n for Jacobi expansions on $[0, \pi]$ that satisfy the property

$$|\phi_n(t)| \leq Cn^{2\max\{\alpha,\beta\}+2}\exp(-cnt^2).$$

Similar results can be also found in the article [66] of Petrushev and Xu.

Space-frequency localization of the Christoffel-Darboux kernel. For the general Jacobi case $\alpha, \beta > -1$, Theorem 3.15 constitutes a new result. For the Chebyshev case $\alpha = \beta = -\frac{1}{2}$, the formulas of Theorem 3.15 are shown in [68]. Similar results for the frequency variance and the uncertainty product of various trigonometric polynomials and wavelets generating a multiresolution analysis on the unit circle can be found in [63], [67], [69], [70] and [78].

Space-frequency localization of the de La Vallée Poussin kernel. Theorem 3.16 on the space-frequency localization of the de La Vallée Poussin kernel is a slightly more general version of [27, Theorem 2.2] in which the ultraspherical case was shown.

Monotonicity of the extremal zeros of orthogonal polynomials. The monotonicity result of Theorem 3.17 is based on the Hellmann-Feynman Theorem and is a variant of the monotonicity results proven by Ismail in [41]. Many similar results based on the Hellmann-Feynman approach, including extremal zeros of Laguerre and of birth and death process polynomials, can be found in [42, Sections 7.3,7.4], [61] and the references therein. The results on the monotonicity of the extremal zeros of the Jacobi polynomials, i.e., Corollary 3.18, Corollary 3.20 and Theorem 3.21 have evolved from joint work with Ferenc Toókos and are novel in this thesis. Corollary 3.19 is due to [81].

Compact two-point homogeneous spaces. For a good introduction into Lie groups, symmetric spaces, two-point homogeneous spaces and the harmonic analysis on these spaces, we refer to the books [34], [36] and [37] of Helgason. Many technical details of compact two-point homogeneous spaces in Section 3.3.1 are also taken from the books [3], [4] and the articles [1], [24], [35] and [84]. Further, a good introduction into spherical harmonics on the unit sphere is the book [60] of Müller. The basis system for the L^2-space on the projective spaces in Proposition 3.22 is taken from the article [80] of Sherman. Related basis systems can be also found in the article [46] and in the book [14].

Optimally space localized spherical polynomials. Optimally space localized spherical polynomials and wavelets on the unit sphere $\mathbb{S}^d$, $d \geq 2$, in combination with associated ultraspherical polynomials were intensively studied by Laín Fernández, in [48] and [47]. In particular, the part of Theorem 3.28 concerning optimally space localized spherical polynomials and wavelets on the unit sphere $\mathbb{S}^d$ was firstly proven in [48].
New in Section 3.3 is the method of the proof based on the monotonicity results of Section 3.2 and the formulas for the optimally space localized polynomials on the projective spaces. Furthermore, the interesting polynomial spaces $\Pi_{\mathcal{R},n}^M$ and the respective optimally space localized polynomials $\mathcal{P}_{\mathcal{R},n}^M$ are discussed here for the first time.

$\mathcal{P}_{12}^{\mathbb{S}^2}(q).$ $\mathcal{P}_{24}^{\mathbb{S}^2}(q).$ $\mathcal{P}_{48}^{\mathbb{S}^2}(q).$

$K_{12}^{\mathbb{S}^2}(q).$ $K_{24}^{\mathbb{S}^2}(q).$ $K_{48}^{\mathbb{S}^2}(q).$

$V_{12}^{\mathbb{S}^2}(q).$ $V_{24}^{\mathbb{S}^2}(q).$ $V_{48}^{\mathbb{S}^2}(q).$

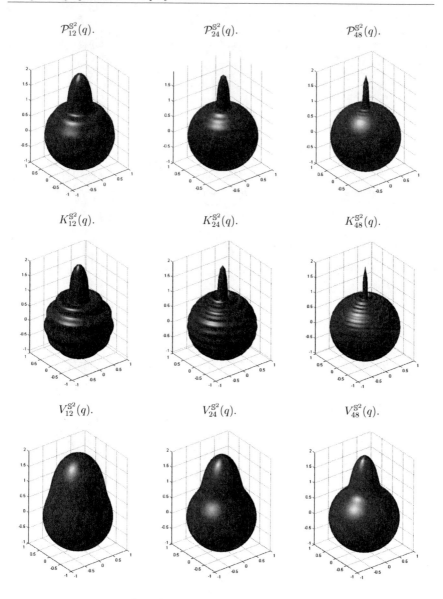

Figure 17: The kernels $\mathcal{P}_n^{\mathbb{S}^2}$, $K_n^{\mathbb{S}^2}$ and $V_n^{\mathbb{S}^2}$ on the unit sphere $\mathbb{S}^2$ centered at the north pole p and normalized such that $\mathcal{P}_n^{\mathbb{S}^2}(p) = \mathcal{P}_{m,n}^{\mathbb{S}^2}(p) = \mathcal{P}_{\mathcal{R},n}^{\mathbb{S}^2}(p) = 1$.

A brief introduction to Riemannian manifolds

In this short appendix, we summarize some basic facts about Riemannian manifolds and introduce the necessary notation for Chapter 2. The details can be found among other standard references in [2], [3], [5], [9], [12], [23], [30] and [36].

A.1. Basic definitions

A *differentiable manifold* M of dimension d is a Hausdorff topological space (with a countable basis) together with a family of injective mappings $\mathbf{x}_i : U_i \subset \mathbb{R}^d \to M$ (U_i open) such that:

(1) $\bigcup_i \mathbf{x}_i(U_i) = M$.

(2) For any pair i, j with $\mathbf{x}_i(U_i) \cap \mathbf{x}_j(U_j) = W \neq \emptyset$, the sets $\mathbf{x}_i^{-1}(W)$ and $\mathbf{x}_j^{-1}(W)$ are open sets in $\mathbb{R}^d$ and $\mathbf{x}_j^{-1}\mathbf{x}_i \in C^\infty$.

(3) The family $\{(U_i, \mathbf{x}_i)\}$ is maximal with respect to (1) and (2).

The pair $(U_i, \mathbf{x}_i)$ with $p \in \mathbf{x}_i(U_i)$ is called a *parametrization* (or a system of coordinates) of the manifold M at p, the set $\mathbf{x}_i(U_i)$ is called a coordinate neighborhood at the point p. A family $\{(U_i, \mathbf{x}_i)\}$ satisfying (1) and (2) is called a *differentiable structure* on M.

On a differentiable manifold M, we consider now differentiable curves $\gamma : \mathbb{R} \supset I \to M$ such that $0 \in I$ and $\gamma(0) = p \in M$. The *tangent vector* at p to the curve γ at $t = 0$ is a function $\gamma'(0)$ that associates to every differentiable function f the value

$$\gamma'(0)f := \left.\frac{d(f \circ \gamma)}{dt}\right|_{t=0}.$$

If we choose a parametrization $\mathbf{x} : U \to M$ at the point $p = \mathbf{x}(0)$, the function f and the curve γ can be written as

$$f \circ \mathbf{x}(q) = f(x_1, \ldots, x_d), \quad q = (x_1, ..., x_d) \in U,$$

and

$$\mathbf{x}^{-1} \circ \gamma(t) = (x_1(t), \ldots, x_d(t)),$$

respectively. So, restricting f to the curve γ, we obtain

$$\gamma'(0)f = \sum_{k=1}^{d} x_k'(0)\left(\frac{\partial f}{\partial x_k}\right)_{x_k=0} = \left(\sum_{k=1}^{d} x_k'(0)\frac{\partial}{\partial x_k}\right)f, \tag{A.1}$$

where $\frac{\partial}{\partial x_k}$ corresponds to the tangent vector of the coordinate curve

$$x_k \to \mathbf{x}(0, \ldots, 0, x_k, 0, \ldots, 0).$$

Equation (A.1) shows that the tangent vector to the curve γ at p depends only on the derivative of γ in a coordinate system and that the set T_pM of all tangent vectors forms a d-dimensional vector space with basis $\left\{\frac{\partial}{\partial x_1}, \ldots, \frac{\partial}{\partial x_d}\right\}$. The vector space T_pM is called *tangent space* of M at p and the set $TM = \{(p,v) : p \in M, \xi \in T_pM\}$ the *tangent bundle* of M.

Now, let M and N be differentiable manifolds and $\varphi : M \to N$ a differentiable mapping. For every $p \in M$ and $\xi \in T_pM$, one can choose a differentiable curve γ on M with $\gamma(0) = p$ and $\gamma'(0) = \xi$. Then, the mapping

$$d\varphi_p : T_pM \to T_{\varphi(p)}N, \quad d\varphi_p(\xi) = (\varphi \circ \gamma)'(0) \tag{A.2}$$

is a linear mapping that does not depend on the choice of γ (see [12], Chapter 0, Proposition 1.2.7). The linear mapping $d\varphi_p$ is called the *differential* of φ at p. If $U \subset M$ and $V \subset N$ are open subsets of the differentiable manifolds M and N, respectively, a mapping $\varphi : U \to V$ is called a *diffeomorphism* if it is differentiable, bijective, and its inverse φ^{-1} is also differentiable. The Inverse Function Theorem implies that if the differential $d\varphi_p$ is an isomorphism from T_pM to $T_{\varphi(p)}N$, then φ is a diffeomorphism from an open neighborhood of p onto an open neighborhood of $\varphi(p)$.

A *vector field* X on a differentiable manifold M is a mapping of M into the tangent bundle TM such that, for any $p \in M$, $X(p) \in T_pM$. The field is called differentiable if the mapping $X : M \to TM$ is of class C^∞. The set of all differentiable vector fields on M is denoted by $\Gamma(TM)$. A vector field V along a curve $\gamma : I \subset \mathbb{R} \to M$ is a differentiable curve $V : I \to TM$ such that $V(t) \in T_{\gamma(t)}M$ for all $t \in I$. The vector field $\gamma'(t) := d\gamma(\frac{d}{dt})$ along γ is called the velocity field (or tangent vector field) of γ.

A *Riemannian metric* on a differentiable manifold M is a correspondence which associates to each point $p \in M$ an inner product $\langle \cdot, \cdot \rangle_p$ on the tangent space T_pM which varies differentiably in the following sense: If $\mathbf{x} : U \subset \mathbb{R}^d \to M$ is a chart around p, with $\mathbf{x}(x_1, \ldots, x_d) = q \in \mathbf{x}(U)$ and the differential $\frac{\partial}{\partial x_k}(q) = d\mathbf{x}_q(0, \ldots, 1, \ldots, 0)$, then $g_{j,k}(x_1, \ldots, x_d) = \langle \frac{\partial}{\partial x_j}(q), \frac{\partial}{\partial x_k}(q) \rangle_q$ are differentiable functions on U. A differentiable manifold endowed with a Riemannian metric is called a *Riemannian manifold*. For the functions $g_{j,k}$, we define the matrix

$$G_{U,\mathbf{x}} = \left[g_{j,k}(x_1, \ldots, x_d) \right]_{j,k=1}^d = \left[\left\langle \frac{\partial}{\partial x_j}(q), \frac{\partial}{\partial x_k}(q) \right\rangle_q \right]_{j,k=1}^d. \tag{A.3}$$

The inverse of $G_{U,\mathbf{x}}$ and its matrix entries are denoted by

$$G_{U,\mathbf{x}}^{-1} = \left[g^{j,k}(x_1, \ldots, x_d) \right]_{j,k=1}^d. \tag{A.4}$$

If M and N are Riemannian manifolds, a diffeomorphism $\varphi : M \to N$ is called an *isometry* if

$$\langle \xi_1, \xi_2 \rangle_p = \langle d\varphi_p(\xi_1), d\varphi_p(\xi_2) \rangle_{\varphi(p)}, \quad \text{for all } p \in M, \ \xi_1, \xi_2 \in T_pM. \tag{A.5}$$

Let $\varphi : M \to N$ be an *immersion*, i.e. φ is differentiable and $d\varphi_p : T_pM \to T_{\varphi(p)}N$ is injective for all $p \in M$. If N has a Riemannian structure, then φ induces a Riemannian structure on M by

$$\langle \xi_1, \xi_2 \rangle_p = \langle d\varphi_p(\xi_1), d\varphi_p(\xi_2) \rangle_{\varphi(p)}.$$

Since $d\varphi_p$ is injective, $\langle \cdot, \cdot \rangle_p$ is positive definite. This metric on M is called the metric induced by φ.

A differentiable map $\varphi : M \to N$ is called a *Riemannian covering map* if $\varphi(M)$ covers N and φ satisfies the isometry condition (A.5) locally. If G is a discrete, free and proper group of isometries on M, then the quotient manifold $N = M/G$ can be endowed with a unique Riemannian metric such that the canonical projection $\varphi : M \to N$ is a Riemannian covering map (see [23, Proposition 2.20]).

On the other hand, if G is a Lie group of isometries on M acting smoothly, properly and freely on M, then the quotient manifold $N = M/G$ can be endowed with a unique Riemannian metric such that the projection map $\varphi : M \to N$ is a *submersion*, i.e., φ is differentiable and $d\varphi_p : T_pM \to T_{\varphi(p)}N$ is an epimorphism for all $p \in M$ (see [23, Proposition 2.28]). Moreover, in this case the map $\varphi : M \to N$ is a smooth *fibration* with fiber G (see [23, Theorem 1.95]), i.e., φ is surjective and there exists an open cover $\{U_i\}_{i \in I}$ of N and diffeomorphisms

$$h_i : \varphi^{-1}(U_i) \rightarrow U_i \times G$$

such that $h_i(\varphi^{-1}(p)) = \{p\} \times G$ for $p \in U_i$.

A.2. Connections and the covariant derivative

Whereas the differentiation of functions is well defined for differentiable manifolds, there is no natural concept of differentiation of vector fields on a manifold M. Therefore, one considers all possibilities of such a differentiation, the so called connections.

A *connection* ∇ on a differentiable manifold M is a mapping $\nabla : \Gamma(TM) \times \Gamma(TM) \rightarrow \Gamma(TM)$ denoted by $(X, Y) \overset{\nabla}{\rightarrow} \nabla_X(Y)$ which satisfies the properties

(1) $\nabla_{fX+gY} Z = f\nabla_X Z + g\nabla_Y Z$,

(2) $\nabla_X(Y + Z) = \nabla_X Y + \nabla_X Z$,

(3) $\nabla_X(fY) = f\nabla_X Y + X(f)Y$,

for $X, Y, Z \in \Gamma(TM)$ and $f, g \in C^\infty(M)$. For any connection ∇, there exists a unique differentiation operator $\frac{D}{dt}$ (cf. [12, Proposition 2.2.2]) defined on the vector space of vector fields along a differentiable curve γ such that:

(1) $\frac{D}{dt}(V + W) = \frac{D}{dt}V + \frac{D}{dt}W$ for vector fields V, W along γ,

(2) $\frac{D}{dt}(fV) = f'V + f\frac{D}{dt}V$ for a differentiable function f,

(3) If V is induced by a vector field Y, i.e. $V(t) = Y(\gamma(t))$, then $\frac{D}{dt}V = \nabla_{\gamma'} Y$.

$\frac{D}{dt}V$ is called the *covariant derivative* of V along the curve γ. A vector field V along a curve $\gamma : I \rightarrow M$ is called *parallel* if $\frac{D}{dt}V = 0$, for all $t \in I$.

On a general differentiable manifold, there exists no connection with distinguished properties. However, if M is a Riemannian manifold with Riemannian metric $\langle \cdot, \cdot \rangle_p$, then there exists a unique connection ∇ on M (cf. [12, Theorem 3.6]) satisfying the following conditions:

(a) ∇ is symmetric, i.e. $\nabla_X Y - \nabla_Y X = [X, Y]$ for all vector fields $X, Y \in \Gamma(TM)$.

(b) ∇ is compatible with the Riemannian metric, i.e. for any smooth curve γ and any pair of parallel vector fields V and W along γ, we have $\langle V, W \rangle_{c(t)} = \text{constant}$.

This canonical connection ∇ of a Riemannian manifold is called *Levi-Civita connection*. If two vector fields X and Y have the representations

$$X = \sum_{i=1}^{d} X_i \frac{\partial}{\partial x_i}, \quad Y = \sum_{i=1} Y_i \frac{\partial}{\partial x_i}$$

in a local chart $\mathbf{x} : U \subset \mathbb{R}^d \to M$ around $p \in M$, then the Levi-Civita connection ∇ can be written in local coordinates as [23, Proposition 2.54],

$$\nabla_X Y = \sum_{i=1}^{d} \left(\sum_{j=1}^{d} X_j \frac{\partial Y_j}{\partial x_j} + \sum_{j,k=1}^{d} \Gamma_{j,k}^i X_j Y_k \right) \frac{\partial}{\partial x_i}, \tag{A.6}$$

where the *Christoffel symbols* $\Gamma_{j,k}^i$ are defined by the relation $\nabla_{\frac{\partial}{\partial x_j}} \frac{\partial}{\partial x_k} = \sum_{i=1}^{d} \Gamma_{j,k}^i \frac{\partial}{\partial x_i}$. Further, we have

$$\Gamma_{j,k}^i = \frac{1}{2} \sum_{l=1}^{d} g^{il} \left(\frac{\partial}{\partial x_j} g_{kl} + \frac{\partial}{\partial x_k} g_{lj} - \frac{\partial}{\partial x_l} g_{jk} \right). \tag{A.7}$$

A.3. Geodesics and the metric space structure

In what follows, we will always assume that M is a Riemannian manifold endowed with the Levi-Civita connection ∇ and that $\frac{D}{dt}$ is the covariant derivative associated to the Levi-Civita connection.

A parametrized curve $\gamma : I \to M$ is called a *geodesic* if the acceleration vector field $\frac{D}{dt}\gamma'$ is zero for every $t \in I$. Due to equation (A.6), in a local chart $\mathbf{x} : U \subset \mathbb{R}^d \to M$, the geodesics are the solutions of the differential equation

$$\frac{d^2 x_i}{dt^2} + \sum_{j,k=1}^{d} \Gamma_{j,k}^i(\mathbf{x}(t)) \frac{dx_k}{dt} \frac{dx_j}{dt} = 0, \quad 1 \le i \le d, \tag{A.8}$$

where $\mathbf{x}(t) = (x_1(t), \dots, x_d(t))$. If $\gamma_\xi(t)$ is a geodesic with initial conditions $\gamma(0) = p$ and $\gamma'(0) = \xi$, then there exists a neighborhood U around p in which $\gamma_\xi(t)$ is uniquely determined and depends smoothly on the parameters t and ξ (see [23, Corollary 2.85]).

If I_ξ is the maximal interval on which $\gamma_\xi(t)$ is defined, then for any $\alpha \in \mathbb{R} \setminus \{0\}$, we have

$$I_{\alpha\xi} = \frac{1}{\alpha} I_\xi, \quad \gamma_{\alpha\xi}(t) = \gamma_\xi(\alpha t).$$

If we denote by $\mathcal{T}_p M$ the subset

$$\mathcal{T}_p M := \{ \xi \in T_p M : 1 \in I_\xi \} \subset T_p M, \tag{A.9}$$

of the tangent space and by $\mathcal{T}M$ the respective subset of the tangent bundle TM, then we can define the so called *exponential map*

$$\exp_p : T_pM \to M, \quad \exp_p(\xi) := \gamma_\xi(1). \tag{A.10}$$

Applied to the whole bundle $\mathcal{T}M$, we denote the exponential map by exp.

By the Riemannian metric structure of M, the length of an element ξ in the tangent space T_pM is given by

$$|\xi| := \langle \xi, \xi \rangle_p^{1/2}.$$

Moreover, we can define the length of a piecewise differentiable curve $\gamma : I \to M$ as

$$l(\gamma) := \int_I |\gamma'(t)| dt. \tag{A.11}$$

In particular, if γ is a geodesic, then

$$\frac{d}{dt} \langle \gamma', \gamma' \rangle = 2 \langle \frac{D}{dt} \gamma', \gamma' \rangle = 0.$$

Thus, the length of the tangent vector γ' is constant. If the geodesic γ is starting at $\gamma(0) = p$, the length of γ from p to $\gamma(t)$ is given by

$$l(\gamma) = \int_0^t |\gamma'(t)| dt = ct,$$

where c denotes the constant length $|\gamma'|$ of γ'. Therefore, the parameter of a geodesic is proportional to the arc length $l(\gamma)$.

With help of the arc length (A.11) of curves, a distance metric $d(p,q)$ between two points p and q can be introduced on M by

$$d(p,q) := \inf_\gamma \int_a^b |\gamma'(t)| dt, \tag{A.12}$$

where γ ranges over all piecewise differentiable paths $\gamma : [a,b] \to M$ satisfying $\gamma(a) = p$ and $\gamma(b) = q$. For $p \in M$ and $\delta > 0$, the open ball and the sphere with center p on M are defined as

$$B(p,\delta) := \{x \in M, \, d(x,p) < \delta\}, \tag{A.13}$$
$$S(p,\delta) := \{x \in M, \, d(x,p) = \delta\}. \tag{A.14}$$

By the same token, we define on the tangent space T_pM

$$\mathfrak{B}(p,\delta) := \{\xi \in T_pM, \, |\xi| < \delta\}, \tag{A.15}$$
$$\mathfrak{S}(p,\delta) := \{\xi \in T_pM, \, |\xi| = \delta\}, \tag{A.16}$$
$$\mathfrak{S}_p := \mathfrak{S}(p,1). \tag{A.17}$$

The distance metric $d : M \times M \to R$ given in equation (A.12) turns M into a metric space (cf. [9, Corollary I.6.1]). Moreover, for $p \in M$ and $\delta > 0$ small enough, we have

$$\exp_p \mathfrak{B}(p, \delta) = B(p, \delta),$$
$$\exp_p \mathfrak{S}(p, \delta) = S(p, \delta),$$

diffeomorphically. Thus, the topology induced by the distance metric $d(\cdot, \cdot)$ coincides with the original topology of M. Moreover, the following local result holds (cf. [23, Theorem 2.92]):

Theorem A.1.
For each $p \in M$, there exists a neighborhood U of p and $\epsilon > 0$ such that any two points q_1 and q_2 of U are joined by a unique geodesic γ in M of length less than ϵ. Moreover, the geodesic γ depends differentiably on its endpoints q_1 and q_2 and its length is given by $l(\gamma) = d(q_1, q_2)$.

A curve γ connecting two points $\gamma(a)$ and $\gamma(b)$ in M is called *minimal* if $l(\gamma|_{[a,b]}) = d(\gamma(a), \gamma(b))$. Thus, by Theorem A.1, a geodesic γ is locally minimal. On the other hand, if a curve γ is parameterized proportional to arc length and γ is locally minimal, then γ is a geodesic (see [23, Corollary 2.94]).

We say that a Riemannian manifold M is *geodesically complete* if for every $p \in M$ and $\xi \in T_pM$, the geodesic $\gamma_\xi(t)$ is defined for all values $t \in \mathbb{R}$, that is, if the exponential map $\exp_p$ is defined on the whole tangent space T_pM. The connection between geodesic completeness and completeness as a metric space is established in the Theorem of Hopf and Rinow (cf. [12, Chapter 7, Theorem 2.8]).

Theorem A.2 (Hopf and Rinow).
Let M be a Riemannian manifold. For $p \in M$, the following assertions are equivalent:

(a) $\exp_p$ is defined on all of T_pM, in particular $\mathcal{T}_pM = T_pM$.

(b) The closed and bounded sets of M are compact.

(c) M is complete as a metric space.

(d) M is geodesically complete.

(e) There exists a sequence of compact subsets $K_n \subset M$, $K_n \subset K_{n+1}$ and $\bigcup_{n=1}^{\infty} K_n = M$, such that if $q_n \notin K_n$ then $d(p, qn) \to \infty$.

In addition, any of the statements above implies that

(f) For any point $q \in M$, there exists a geodesic γ joining p to q with minimal length $l(\gamma) = d(p, q)$.

Note that the geodesic in part (f) of Theorem A.2 is in general not unique.

A.4. The cut locus

For $p \in M$ and a unit vector $\xi \in \mathfrak{S}_p \subset T_pM$, we define the distance $R(\xi)$ to the cut point of p along the geodesic $\gamma_\xi(t)$ by

$$R(\xi) := \sup_{t > 0} \left\{ t\xi \in T_pM : \ d(p, \gamma_\xi(t)) = t \right\}. \tag{A.18}$$

In other words, $R(\xi)$ is the maximal distance in direction ξ for which the exponential map $\exp_p$ is isometric. The point $\gamma_\xi(R(\xi))$ is called the *cut point* of p along the geodesic $\gamma_\xi(t)$. The geodesic $\gamma_\xi(t)$ minimizes the distance between p and $\exp_p(t\xi)$ for all $t \in [0, R(\xi))$, and fails to minimize the distance for all $t > R(\xi)$. Indeed, if there exists a $t \in [0, R(\xi))$ such that $d(p, \exp_p(t\xi)) < t$, then the triangle inequality implies

$$d(p, R(\xi)) \leq d(p, \exp_p(t\xi)) + d(\exp_p(t\xi), R(\xi)) < t + (R(\xi) - t) = R(\xi),$$

a contradiction. If $t < R(\xi)$, then γ_ξ is the only minimal geodesic between p and $\gamma_\xi(t)$. Moreover, if $R(\xi)$ is finite and $R(\xi)\xi \in T_pM$, then γ_ξ minimizes also the distance between p and $\exp_p(R(\xi)\xi)$.

We consider now R as a function on the unit sphere $\mathfrak{S}_p$. The following theorem collects some information on the smoothness of R.

Theorem A.3.
For $p \in M$, the function $R : \mathfrak{S}_p \to (0, \infty]$ has the following properties:

(a) *R is upper semicontinuous on $\mathfrak{S}_p$ [9, Theorem III.2.1].*

(b) *If M is geodesically complete, then R is continuous on $\mathfrak{S}_p$ [9, Theorem III.2.1].*

(c) *If M is a compact real analytic Riemannian manifold of dimension d, then the surface $\xi \in \mathfrak{S}_p \to R(\xi)\xi$ is a $(d-1)$-dimensional simplicial complex [7].*

(d) *If M is compact, then the distance function R is a Lipschitz continuous function on $\mathfrak{S}_p$ [43].*

For $p \in M$, we define the *tangential cut locus* $\mathfrak{C}_p$ in the tangent space T_pM by

$$\mathfrak{C}_p := \{R(\xi)\xi : \ R(\xi) < \infty, \ \xi \in \mathfrak{S}_p\} \cap T_pM, \tag{A.19}$$

and the *cut locus C_p* of p in M, by

$$C_p := \exp_p \mathfrak{C}_p. \tag{A.20}$$

Moreover, we set

$$\mathfrak{D}_p := \{t\xi : \ 0 \leq t < R(\xi), \ \xi \in \mathfrak{S}_p\}, \tag{A.21}$$

and

$$D_p := \exp_p \mathfrak{D}_p. \tag{A.22}$$

If M is complete, we get the following decomposition (cf. [23, Proposition 2.113 and Corollary 3.77])

Theorem A.4.
Let M be a complete Riemannian manifold. Then, for any $p \in M$ we have the disjoint decomposition

$$M = D_p \cup C_p. \tag{A.23}$$

Moreover, $\mathfrak{D}_p$ is the largest star-shaped open domain in T_pM (with respect to the origin) for which the exponential map is a diffeomorphism with $D_p = \exp_p(\mathfrak{D}_p) = M \setminus C_p$.

For $p \in M$, we define the *injectivity radius* $\operatorname{inj} p$ of p by

$$\operatorname{inj} p := \inf_{\xi \in \mathfrak{S}_p} \{R(\xi)\} \tag{A.24}$$

and the injectivity radius of M by

$$\operatorname{inj} M := \inf_{p \in M} \{\operatorname{inj} p\}. \tag{A.25}$$

A.5. Integration on Riemannian manifolds

Let $\mathbf{x} : U \subset \mathbb{R}^d \to M$ be a chart on M. For each $q \in \mathbf{x}(U)$, we consider the matrices $G_{U,\mathbf{x}}$ defined in (A.3). Then, the determinant $\det G_{U,\mathbf{x}}$ is positive and, on U, we can define the positive measure $\sqrt{\det G_{U,\mathbf{x}}} dx_1 \cdots dx_d$. Thus, by $\mathbf{x}(\sqrt{\det G_{U,\mathbf{x}}} dx_1 \cdots dx_d)$ we get a positive measure on $\mathbf{x}(U) \subset M$ that is independent from the particular choice of the chart $\mathbf{x}$ (see [9, Section III.3]). By a partition of unity argument we construct now a global Riemannian measure. We take an atlas

$$\left\{\mathbf{x}_i : U_i \subset \mathbb{R}^d \to M, \ i \in I\right\}$$

on M, a subordinate partition of unity $\{\phi_i : \ i \in I\}$, and define the global Riemannian measure μ_M by

$$d\mu_M := \sum_{i \in I} \phi_i \, \mathbf{x}_i(\sqrt{\det G_{U_i,\mathbf{x}_i}} dx_1^i \cdots dx_d^i). \tag{A.26}$$

The measure μ_M is positive and well-defined (cf. [77, Chapter IV, Theorem 17]), i.e. independent of the choice of atlas and the subordinate partition of unity. Further, a function f is measurable with respect to $d\mu_M$ if and only if $f \circ \mathbf{x}_i$ is measurable on U_i for any chart $\mathbf{x}_i : U_i \subset \mathbb{R}^d \to M$. The measure μ_M is called the *canonical measure* or the *Riemannian measure* on M.

Now, we consider the measure on M induced by the exponential map. Let therefore $p \in M$ and V and U be open neighborhoods of $0 \in T_pM$ and of p in M, respectively, such that the exponential $\exp_p|_V$ is a diffeomorphism from V onto U. We identify the tangent space at $\xi \in T_pM$ with T_pM itself and consider the differential

$$(d\exp_p)_\xi : T_\xi T_pM \to T_{\exp_p(\xi)}M. \tag{A.27}$$

The differential $(d\exp_p)_\xi$ is a linear mapping and can therefore be considered as an isomorphism from the Euclidean space T_pM onto the Euclidean space $T_{\exp_p(\xi)}M$. We define

$$\theta(\xi) := \det((d\exp_p)_\xi). \tag{A.28}$$

and denote by $d\xi$ the standard Lebesgue measure on the Euclidean space T_pM. Then, for any integrable function f on U, we have (cf. [3, Proposition C.III.2])

$$\int_U f(q)d\mu_M(q) = \int_V f(\exp_p(\xi))\theta(\xi)d\xi. \tag{A.29}$$

From Theorem A.4, we know that D_p is the largest open subset of M for which the exponential map is a diffeomorphism. If M is a complete Riemannian manifold, then the cut locus $C_p = M \setminus D_p$ is a set of Riemannian measure zero [9, Proposition III.3.1], and we can deduce the formula

$$\int_M f(q)d\mu_M(q) = \int_{\mathfrak{D}_p} f(\exp_p(\xi))\theta(\xi)d\xi \tag{A.30}$$

for the integral of a function f over the whole manifold M.

A.6. Curvature

The *curvature* R of a Riemannian manifold M is a correspondence that associates to every pair $X, Y \in \Gamma(TM)$ of vector fields a mapping $R(X,Y) : \Gamma(TM) \to \Gamma(TM)$ given by

$$R(X,Y)Z := \nabla_Y \nabla_X Z - \nabla_X \nabla_Y Z + \nabla_{[X,Y]}Z, \quad Z \in \Gamma(TM), \tag{A.31}$$

where ∇ is the usual Levi-Civita-connection on M. The curvature tensor R has the following properties (cf. [12], Chapter 4, Proposition 2.2 and 2.4)

(i) R is bilinear in $\Gamma(TM) \times \Gamma(TM)$, i.e.,

$$R(fX_1 + gX_2, Y_1) = fR(X_1, Y_1) + gR(X_2, Y_1),$$
$$R(X_1, fY_1 + gY_2) = fR(X_1, Y_1) + gR(X_1, Y_2),$$
$$f, g \in C^\infty(M), \quad X_1, X_2, Y_1, Y_2 \in \Gamma(TM).$$

(ii) For $X, Y \in \Gamma(TM)$, the curvature operator $R(X,Y) : \Gamma(TM) \to \Gamma(TM)$ is linear, i.e.,

$$R(X,Y)(Z + W) = R(X,Y)Z + R(X,Y)W,$$
$$R(X,Y)(fZ) = fR(X,Y)Z, \quad f \in C^\infty(M), \ Z, W \in \Gamma(TM).$$

(iii) $R(X,Y)Z + R(Y,Z)X + R(Z,X)Y = 0$ (Bianchi identity).

For $p \in M$, we consider a two-dimensional subspace of T_pM spanned by the vectors $\xi_1, \xi_2 \in T_pM$. Then, the real value

$$K(\xi_1, \xi_2) := \frac{\langle R(\xi_1, \xi_2)\xi_1, \xi_2 \rangle_p}{\sqrt{|\xi_1|^2|\xi_2|^2 - \langle \xi_1, \xi_2 \rangle_p}} \tag{A.32}$$

is known as the sectional curvature of the space spanned by the vectors $\xi_1, \xi_2 \in T_pM$ at p. The value $K(\xi_1, \xi_2)$ does not depend on the particular choice of the vectors ξ_1 and ξ_2, see [12, chapter 4, Proposition 3.1]. Certain averages of the sectional curvature are known as *Ricci curvature* and *scalar curvature*. In particular, if $\{e_1, e_2, \cdots, e_d\}$ denotes an orthonormal basis of the tangent space T_pM, then the Ricci curvature tensor $Ric : T_pM \times T_pM \rightarrow \mathbb{R}$ is defined as

$$Ric(\xi_1, \xi_2) := \sum_{i=1}^{d} \langle R(\xi_1, e_i)\xi_2, e_i \rangle_p, \tag{A.33}$$

and the scalar curvature as

$$K := \sum_{i,j=1, i \neq j}^{d} \langle R(e_i, e_j)e_i, e_j \rangle_p. \tag{A.34}$$

The Ricci and the scalar curvature are independent of the particular choice of the orthonormal basis.

A.7. The Laplace-Beltrami operator

For any differentiable function f on the Riemannian manifold M, the gradient grad f of f is the vector field defined by

$$\langle grad f(p), \xi \rangle_p := df_p(\xi), \quad \text{for all } p \in M, \xi \in T_pM. \tag{A.35}$$

Further, for a differentiable vector field X on M, the divergence of X, div $X : M \rightarrow \mathbb{R}$, is defined by

$$div X(p) := tr(\xi \rightarrow \nabla_\xi X), \tag{A.36}$$

where ξ ranges over all tangent vectors in T_pM and ∇ denotes as usual the Levi-Civita connection on M. If $\mathbf{x} : U \subset \mathbb{R}^d \rightarrow M$ is a chart on M, then grad f can be written in local coordinates as

$$grad f = \sum_{j,k=1}^{d} \frac{\partial(f \circ \mathbf{x})}{\partial x_j} g^{jk} \frac{\partial}{\partial x_k}. \tag{A.37}$$

Moreover, one has for differentiable functions f and g,

$$\text{grad}(f + g) = \text{grad}\, f + \text{grad}\, g,$$
$$\text{grad}(fg) = g\,\text{grad}\, f + f\,\text{grad}\, g.$$

If the vector field X has the representation $X = \sum_{i=1}^{d} X_i \frac{\partial}{\partial x_i}$, then the divergence $\text{div}\, X$ can be written in the local coordinates as

$$\text{div}\, f = \frac{1}{\sqrt{\det G_{U,\mathbf{x}}}} \sum_{i=1}^{d} \frac{\partial(X_i \sqrt{\det G_{U,\mathbf{x}}})}{\partial x_j}. \tag{A.38}$$

Further, for a differentiable function f and differentiable vector fields X and Y, the following properties hold:

$$\text{div}(X + Y) = \text{div}\, X + \text{div}\, Y,$$
$$\text{div}(fX)(p) = f(p)\,\text{div}\, X(p) + \langle \text{grad}\, f, X \rangle_p.$$

For a C^2-function f on M, we define the Laplacian of f by

$$\Delta_M f := \text{div}\,\text{grad}\, f. \tag{A.39}$$

The operator Δ_M is referred to as the Laplace-Beltrami operator. In a local chart $\mathbf{x} : U \subset \mathbb{R}^d \to M$, we have

$$\Delta_M f = \frac{1}{\sqrt{\det G_{U,\mathbf{x}}}} \sum_{j,k=1}^{d} \frac{\partial}{\partial x_j}\left(\sqrt{\det G_{U,\mathbf{x}}}\, g^{jk} \frac{\partial(f \circ \mathbf{x})}{\partial x_k} \right). \tag{A.40}$$

Moreover, the Laplace-Beltrami operator Δ_M satisfies the properties

$$\Delta_M(f + g) = \Delta_M f + \Delta_M g,$$
$$\text{div}(f\,\text{grad}\, g)(p) = f(p)\Delta_M g(p) + \langle \text{grad}\, f, \text{grad}\, g \rangle_p,$$

where f and g are C^2-functions on M.

Basics on function spaces and operators on Hilbert spaces

B.1. Function spaces

First of all, we summarize some basic facts about general L^p-spaces. A detailed elaboration of this topic can be found in the classical monograph [38, Section 13].

Let $1 \leq p \leq \infty$ and $(X, \mathcal{A}, \mu)$ be an arbitrary measure space. Then, we define the spaces

$$\mathcal{L}^p(X) := \left\{ f : X \to \mathbb{C} : f \ \mathcal{A}\text{-measurable}, \int_X |f(x)|^p d\mu(x) < \infty \right\}, \quad 1 \leq p < \infty,$$

$$\mathcal{L}^\infty(X) := \left\{ f : X \to \mathbb{C} : f \ \mathcal{A}\text{-measurable}, \ \underset{x \in X}{\mathrm{ess\,sup}} |f(x)| < \infty \right\}.$$

On $\mathcal{L}^p(X)$, we define the functional $\| \cdot \|_p$ by

$$\|f\|_p := \left(\int_X |f(x)|^p d\mu(x) \right)^{\frac{1}{p}}, \quad 1 \leq p < \infty,$$

$$\|f\|_\infty := \underset{x \in X}{\mathrm{ess\,sup}} |f(x)|.$$

For $f \in \mathcal{L}^p(X)$, the function $f \to \|f\|_p$ satisfies all axioms of a norm except for the positivity condition, i.e. $\|f\|_p > 0$ if $f \neq 0$. Therefore, let

$$\mathcal{N} := \{ f \in \mathcal{L}^p(X) : f = 0 \ \mu\text{-a.e.} \}.$$

Then, $\mathcal{N}$ is a closed linear subset of $\mathcal{L}^p(X)$ and we can define the quotient space

$$L^p(X) := \mathcal{L}^p(X)/\mathcal{N}, \quad 1 \le p \le \infty. \tag{B.1}$$

Now it is straightforward to show that $L^p(X)$ with the norm $\|\cdot\|_p$ is a linear normed vector spaces, where $f = g$ means that $f(x) = g(x)$ for μ-a.e. $x \in X$,i.e., for all $x \in X$ except a set of μ-measure zero. Moreover, $L^p(X)$ with the metric $d(f,g) = \|f - g\|_p$ is a complete metric space and, hence, a Banach space.

The special case $p = 2$ is particularly interesting. In this case, one can define the inner product

$$\langle \cdot, \cdot \rangle : L^2(X) \times L^2(X) \to \mathbb{C}, \quad \langle f, g \rangle = \int_X f(x)\overline{g(x)}d\mu(x), \tag{B.2}$$

that turns $L^2(X)$ into a Hilbert space. Further, the norm on $L^2(X)$ can be expressed as $\|f\|_2 = \sqrt{\langle f, f \rangle}$.

If the measure space $(X, \mathcal{A}, \mu)$ consists of a Riemannian manifold M endowed with the Borel σ-algebra and the Riemannian measure μ_M, we denote the L^p-spaces in (B.1) as $L^p(M)$ and the scalar product in (B.2) as $\langle \cdot, \cdot \rangle_M$.

Next, we consider absolutely continuous functions on an interval $[a, b]$ and on the real line $\mathbb{R}$. For a detailed introduction, we refer to [38, Section 18]. A function $f : [a, b] \to \mathbb{C}$ is called *absolutely continuous* if f admits the representation

$$f(x) - f(a) = \int_a^x f_0(t)dt \tag{B.3}$$

for a function $f_0 \in L^1([a, b])$. The function f_0 is called the *Radon-Nikodym derivative* of f. An absolutely continuous function f is uniformly continuous on $[a, b]$ and differentiable for almost all $t \in (a, b)$. Further, for the pointwise derivative f' we have $f'(t) = f_0(t)$ for a.e. $t \in [a, b]$. Hence, $f' = f_0$ in $L^1([a, b])$ and we can from now on use the symbol f' also for the Radon-Nikodym derivative of f.

The space of all absolutely continuous functions on $[a, b]$ is denoted by $AC([a, b])$. If $f, g \in AC([a, b])$, then also fg is absolutely continuous and for the Radon-Nikodym derivative the usual product formula holds, i.e.

$$(fg)'(t) = f'(t)g(t) + g'(t)f(t)$$

for a.e. $t \in [a, b]$. Moreover, integration by parts for two functions $f, g \in AC([a, b])$ reads as follows (cf. [38, Corollary 18.20]):

$$\int_a^b f'(t)g(t)dt = f(a)g(a) - f(b)g(b) - \int_a^b f(t)g'(t)dt. \tag{B.4}$$

If the underlying set is the real line $\mathbb{R}$, we define the space of locally absolutely continuous functions $AC_{loc}(\mathbb{R})$ as

$$AC_{loc}(\mathbb{R}) := \left\{ f : \mathbb{R} \to \mathbb{C} : f|_{[a,b]} \in AC([a, b]), \text{ for all } [a, b] \subset \mathbb{R} \right\}, \tag{B.5}$$

and the space of absolutely continuous functions on $\mathbb{R}$ as

$$AC(\mathbb{R}) := \left\{ f \in AC_{loc}(\mathbb{R}) : \ f' \in L^1(\mathbb{R}) \right\}. \tag{B.6}$$

So, every absolutely continuous function $f \in AC(\mathbb{R})$ can be written as $f(x) = \int_{-\infty}^{x} f_0(t)dt$, where $f_0 \in L^1(\mathbb{R})$ and $f'(t) = f_0(t)$ for a.e. $t \in \mathbb{R}$. Further, if $f, g \in AC(\mathbb{R})$, then the following formula holds (cf. [38, Corollary 18.21]):

$$\int_{-\infty}^{\infty} f'(t)g(t)dt + \int_{-\infty}^{\infty} f(t)g'(t)dt = \lim_{x \to \infty} f(x) \lim_{x \to \infty} g(x). \tag{B.7}$$

Let (X, d_X) and (Y, d_Y) be two metric spaces with metric d_X and d_Y, respectively. Then, a function $f : X \to Y$ is called *Lipschitz continuous* on X if there exists a constant $K \geq 0$ such that for all $x_1, x_2 \in X$ the following inequality holds:

$$d_Y(f(x_1), f(x_2)) \leq K d_X(x_1, x_2).$$

Finally, we list some classical function spaces that are used within the text.

$$C([a,b]) := \{f : [a,b] \to \mathbb{C} : \ \text{f continuous on [a,b]}\}, \tag{B.8}$$
$$C_{2\pi} := \{f \in C([-\pi,\pi]) : \ f(-\pi) = f(\pi)\}, \tag{B.9}$$
$$AC_{2\pi} := \{f \in AC([-\pi,\pi]) : \ f(-\pi) = f(\pi)\}, \tag{B.10}$$
$$C^1([a,b]) := \{f \in C([a,b]) \ \text{differentiable on (a,b)} : \ f' \in C([a,b])\}, \tag{B.11}$$
$$C^k([a,b]) := \left\{f \in C^{k-1}([a,b]) \ \text{differentiable on (a,b)} : \ f^{(k)} \in C([a,b])\right\}. \tag{B.12}$$

B.2. The Stone-Weierstrass Theorem

There exist various versions of the Stone-Weierstrass Theorem. In the following, we will present two of them that are needed within the text.

First, let X denote a nonvoid compact Hausdorff space and $C(X)$ the space of all continuous complex-valued functions on X. Endowed with the norm

$$\|f\|_\infty = \sup_{x \in X} |f(x)|, \quad f \in C(X), \tag{B.13}$$

the space $C(X)$ is a Banach space. Further, if we define an involution operator $\tilde{\ }$ on $C(X)$ by $\tilde{f}(x) := \overline{f(x)}$, the space $C(X)$ with addition and multiplication defined pointwise, and endowed with the involution $\tilde{\ }$ is a commutative C^*-algebra.

We say that a subset $\mathcal{A} \subset C(X)$ is a *separating* family of functions on X if for every $x, y \in X$, $x \neq y$, there exists a function $f \in \mathcal{A}$ such that $f(x) \neq f(y)$. We say that $\mathcal{A}$ is *closed under complex conjugation* if for every $f \in \mathcal{A}$, also the involution $\tilde{f} \in \mathcal{A}$. Then, the following result holds (cf. [38, Theorem 7.34]):

Theorem B.1 (Stone-Weierstrass Theorem).
Let $\mathcal{A}$ be a separating subalgebra of $C(X)$ that contains the constant functions and that is closed under complex conjugation. Then, the functions from $\mathcal{A}$ are dense in $C(X)$ in the topology induced by the norm $\|\cdot\|_\infty$.

There exists also a noncompact version of the Stone-Weierstrass theorem in the case that X is a locally compact Hausdorff space. In this case, we consider the closed subalgebra $C_0(X) \subset C(X)$ of continuous functions f on X with the property that for every $\epsilon > 0$ there exists a compact set $K(\epsilon, f) \subset X$ such that $|f(x)| < \epsilon$ for all $x \in X \setminus K(\epsilon, f)$. A family of functions $\mathcal{A} \subset C_0(X)$ is said to *vanish nowhere* if for every $x \in X$ there exists a function $f \in \mathcal{A}$ such that $f(x) \neq 0$. Now, we get the following result (see [38, Exercise 7.37]):

Theorem B.2 (Stone-Weierstrass Theorem - locally compact version).
Let X be a locally compact Hausdorff space. Let $\mathcal{A}$ be a separating subalgebra of $C_0(X)$ that vanishes nowhere and that is closed under complex conjugation. Then, the functions from $\mathcal{A}$ are dense in $C_0(X)$ in the topology induced by the norm $\|\cdot\|_\infty$.

B.3. Operators on Hilbert spaces

In this last part, we give some basic facts about operators on Hilbert spaces. A detailed introduction can be found in the monographs [76] and [85].

Let $\mathcal{H}$ be a Hilbert space with scalar product $\langle\cdot,\cdot\rangle$. An *operator* $A : \mathcal{H} \supset \mathcal{D}(A) \to \mathcal{H}$ is a linear mapping whose domain of definition $\mathcal{D}(A)$ is a subspace of $\mathcal{H}$.

The operator A is called *densely defined* if $\mathcal{D}(A)$ is a dense subset of $\mathcal{H}$. The operator A is called *bounded* if $\mathcal{D}(A) = \mathcal{H}$ and $\|A\| = \sup_{v \in \mathcal{H}, \|v\|=1} \|Av\| < \infty$. The value $\|A\|$ is then called *operator norm* of A. The operator A is called *closed* if the graph $\mathcal{G}(A) = \{(v, Av) : v \in \mathcal{D}(A)\}$ is a closed subset of $\mathcal{H} \times \mathcal{H}$.
An operator $B : \mathcal{H} \supset \mathcal{D}(B) \to \mathcal{H}$ is called an *extension* of A if $\mathcal{D}(A) \subset \mathcal{D}(B)$ and $Bv = Av$ for all $v \in \mathcal{D}(A)$.

To introduce the *Hilbert space adjoint* A^* of A, we consider the domain

$$\mathcal{D}(A^*) := \{w \in \mathcal{H} : v \to \langle Av, w\rangle \text{ continuous on } \mathcal{D}(A)\}.$$

If $w \in \mathcal{D}(A^*)$, the functional $v \to \langle Av, w\rangle$ can be extended to a continuous linear functional on $\mathcal{H}$ by the Hahn-Banach theorem. Therefore, if $\mathcal{D}(A)$ is dense in $\mathcal{H}$, there exists an unique element $A^*w \in \mathcal{H}$ satisfying

$$\langle Av, w\rangle = \langle v, A^*w\rangle, \quad v \in \mathcal{D}(A).$$

In this way, we introduce the well-defined adjoint operator $A^* : \mathcal{D}(A^*) \to \mathcal{H}$.

The product AB of two operators A and B is naturally defined by $ABv = A(Bv)$ with the domain
$$\mathcal{D}(AB) := \{v \in \mathcal{D}(B) : \ Bv \in \mathcal{D}(A)\}.$$

Further, if A, B and AB are densely defined operators on $\mathcal{H}$, then $(AB)^*$ is an extension of $B^* A^*$.

The operator A is called *symmetric* if

$$\langle Av, w \rangle = \langle v, Aw \rangle$$

for all $v, w \in \mathcal{D}(A)$. In this case the adjoint A^* is an extension of A. If moreover $\mathcal{D}(A) = \mathcal{D}(A^*)$ is satisfied, then A is said to be *self-adjoint*.

A densely defined closed operator A is called *normal* if $A^* A = A A^*$ holds. If A is normal, then $\mathcal{D}(A) = \mathcal{D}(A^*)$ and $\|Av\| = \|A^* v\|$ for all $v \in \mathcal{H}$. Clearly, every self-adjoint operator A is also normal. A further important subclass of normal operators is the class of *unitary* operators. An operator B is called unitary if it is bounded on $\mathcal{H}$ and satisfies the property $B^* B = B B^* = I$. In this case, one has the identity

$$\langle Bv, Bv \rangle = \langle B^* Bv, v \rangle = \langle v, v \rangle$$

and the adjoint B^* corresponds to the inverse operator B^{-1}.

Nomenclature

Chapter 1

$AC_{loc}(\mathbb{R})$	locally absolutely continuous functions on $\mathbb{R}$, p. 12, 153
$AC_{2\pi}$	absolutely continuous 2π-periodic functions on $[-\pi, \pi]$, p. 14, 152
$L^2([0, \pi], w)$	space of weighted square integrable functions on $[0, \pi]$, p. 17
$\tilde{w}$	symmetric extension of the weight function w, p. 18, 27
$L_e^2([-\pi, \pi], \tilde{w})$	space of even, weighted L^2-functions on $[-\pi, \pi]$, p. 18
e	even extension operator, p. 18, 27, 40, 49
r	reduction operator, p. 18, 27, 40, 49
T	Dunkl operator, p. 19, 27
$\varepsilon(f)$	mean value of the function $f \in L^2([0, \pi], w)$, p. 23
$\rho(f)$	integral term on the right hand side of inequality (1.37), p. 23
$\mathrm{var}_S(f)$	position variance of $f \in L^2([0, \pi], w)$, p. 23
$\mathrm{var}_F(f)$	frequency variance of $f \in L^2([0, \pi], w)$, p. 23
$w_{\alpha\beta}$	weight function of the Jacobi polynomials, p. 30
$P_n^{(\alpha,\beta)}$	Jacobi polynomial of degree n, p. 30
$L_{\alpha\beta}$	second order differential operator of the Jacobi polynomials, p. 31
$\varepsilon_{\alpha\beta}(f)$	mean value of the function $f \in L^2([0, \pi], w_{\alpha\beta})$, p. 31

Chapter 2

Z_π^d	d-dimensional cylinder of length π, p. 38
$\mathbb{S}^{d-1}$	$(d-1)$-dimensional unit sphere in $\mathbb{R}^d$, p. 38
μ	standard Riemannian measure on $\mathbb{S}^{d-1}$ and $\mathfrak{S}_p$, p. 38, 56
$C(Z_\pi^d)$	space of continuous functions on Z_π^d, p. 38
$L^2(Z_\pi^d, W)$	Hilbert space of weighted L^2-functions on Z_π^d, p. 39
X^d	doubled d-dimensional cylinder of length 2π, p. 39
$\tilde{W}$	symmetric extension of the weight function W, p. 39
$\check{}$	reflection operator, p. 39
$L_e^2(X^d, \tilde{W})$	space of even, weighted L^2-functions on X^d, p. 39
T^X	Dunkl operator on the Hilbert space $L^2(X^d, \tilde{W})$, p. 40
$C_{2\pi}^{1,t}(X^d)$	space of continuously differentiable functions on X^d in t, p. 41
Z_∞^d	d-dimensional one-sided tube, p. 48
Y^d	d-dimensional two-sided tube, p. 49
M	Riemannian manifold, compact in Section 2.2, p. 52, 141

T_pM	tangential space at the point $p \in M$, p. 52, 140
μ_M	canonical Riemannian measure on M, p. 52, 147
$L^2(M)$	space of square integrable functions on M, p. 53
$d(p,q)$	distance metric between two points $p, q \in M$, p. 53, 144
$B(p,\delta)$	open ball with center p and radius δ on M, p. 53, 144
$S(p,\delta)$	sphere with center p and radius δ on M, p. 53, 144
$\mathfrak{B}(p,\delta)$	open ball centered at 0 with radius δ in T_pM, p. 53, 144
$\mathfrak{S}(p,\delta)$	sphere centered at 0 with radius δ in T_pM, p. 53, 144
$\mathfrak{S}_p$	unit sphere in the tangential space T_pM, p. 53, 144
γ_ξ	geodesic with initial conditions $\gamma_\xi(0) = p$ and $\gamma'_\xi(0) = \xi$, p. 53, 143
$\exp_p$	the exponential map from T_pM to M, p. 53, 144
$\mathfrak{D}_p$	maximal set in T_pM for which $\exp_p$ is diffeomorphic, p. 54, 146
$\mathfrak{C}_p$	tangential cut locus of p, p. 54, 146
D_p	image of $\mathfrak{D}_p$ under $\exp_p$, p. 54, 146
C_p	cut locus of p, p. 54, 146
$\theta(\xi)$	Jacobian determinant of $\exp_p$ at $\xi \in T_pM$, p. 54
P	Polar transform, p. 55
$R(\xi)$	geodesic distance to the cut locus C_p in direction ξ, p. 55, 146
Z_R^d	cylinder with right boundary given by the function R, p. 55
$\exp_p^* f$	pull back of the function f by $\exp_p$, p. 56
f^*	pull back of the function f by P and $\exp_p$, p. 56
$\Theta(t,\xi)$	Jacobian determinant of $\exp_p$ in geodesic polar coordinates, p. 56
L_R	lipeomorphism determined by the distance function R, p. 57
$\mathrm{L}_R^* f^*$	pull back of the function f by the mapping $\exp_p \mathrm{PL}_R$, p. 57
$W_{M,p}(\tau,\xi)$	Jacobian determinant of the mapping $\exp_p \mathrm{PL}_R$ on Z_π^d, p. 57
$\frac{\partial}{\partial t}*$	radial differential operator for functions on M, p. 62
$\mathrm{var}_{F,p}^M(f)$	radial frequency variance of the function f on M, p. 63
$T_{M,p}^X$	Dunkl operator on the Hilbert space $L^2(X^d, \tilde{W}_{M,p})$, p. 63
$\varepsilon_p(f)$	mean value of the function f with respect to $p \in M$, p. 64
$\rho_p(f)$	integral term on the right hand side of inequality (2.90), p. 64
$\mathrm{var}_{S,p}^M(f)$	position variance of the function f on M, p. 64
Δ_M	Laplace-Beltrami operator on the manifold M, p. 66, 149
$\Delta_{p,t}$	radial part of the Laplace-Beltrami operator on M, p. 66
Ω	compact star-shaped subdomain of M, p. 67
$\partial\Omega$	boundary of Ω, p. 67
$Q(\xi)$	geodesic length from p to $\partial\Omega$ in direction $\xi \in \mathfrak{S}_p$, p. 67
E	Riemannian manifold diffeomorphic to $\mathbb{R}^d$, p. 71
$\mathbb{S}_r^d$	sphere with radius r in $\mathbb{R}^{d+1}$, p. 80
$\mathbb{RP}_r^d$	real projective space with diameter $r\pi$, p. 81
$\mathbb{CP}_r^d$	complex projective space with diameter $r\pi$, p. 82
$\mathbb{HP}_r^d$	quaternionic projective space with diameter $r\pi$, p. 82
$\mathbb{C}a_r^d$	Cayley plane with diameter $r\pi$, p. 83
$\mathbb{T}_r^d$	flat torus with diameter $\sqrt{2}r$, p. 85

$\mathbb{H}_r^d$	hyperbolic space with negative curvature $-\frac{1}{r^2}$, p. 86
$K(\xi_1, \xi_2)$	sectional curvature of the space span$\{\xi_1, \xi_2\}$, p. 87, 149
Ric	Ricci curvature tensor on $T_pM \times T_pM$, p. 88, 149

Chapter 3

$p_n^{(\alpha,\beta)}$	orthonormal Jacobi polynomials, p. 94
$\Pi_n^{(\alpha,\beta)}, \Pi_{m,n}^{(\alpha,\beta)}, \Pi_{\mathcal{R},n}^{(\alpha,\beta)}$	polynomial subspaces of $L^2([0,\pi], w_{\alpha\beta})$, p. 94
$\mathbb{S}_n^{(\alpha,\beta)}, \mathbb{S}_{m,n}^{(\alpha,\beta)}, \mathbb{S}_{\mathcal{R},n}^{(\alpha,\beta)}$	unit spheres in the polynomial spaces, p. 94
$\mathcal{P}_n^{(\alpha,\beta)}, \mathcal{P}_{m,n}^{(\alpha,\beta)}, \mathcal{P}_{\mathcal{R},n}^{(\alpha,\beta)}$	optimally space localized polynomials, p. 96
a_l, b_l	coefficients of the three-term recurrence relation of $p_n^{(\alpha,\beta)}$, p. 96
$p_n^{(\alpha,\beta)}(\cdot, c)$	associated Jacobi polynomials, p. 97
$p_n^{(\alpha,\beta)}(\cdot, c, \gamma, \delta)$	scaled co-recursive associated Jacobi polynomials, p. 97
J_n^m	Jacobi matrix corresponding to the polynomial $p_n^{(\alpha,\beta)}(\cdot, m)$, p. 97
$\mathcal{L}_n, \mathcal{L}_n^m, \mathcal{L}_n^{\mathcal{R}}$	subsets of the unit spheres $\mathbb{S}_n^{(\alpha,\beta)}, \mathbb{S}_{m,n}^{(\alpha,\beta)}, \mathbb{S}_{\mathcal{R},n}^{(\alpha,\beta)}$, p. 101
$K_n^{(\alpha,\beta)}$	Christoffel-Darboux kernel, p. 110
V_n	de La Vallée Poussin kernel, p. 114
$Q_n^{(\alpha,\beta)}(x, c)$	monic associated Jacobi polynomials, p. 117
P_l^M	radial spherical polynomial of order l on M, p. 124
$P_{l,k,j}^M$	general spherical polynomial of order l on M, p. 125
$\Pi_n^M, \Pi_{m,n}^M, \Pi_{\mathcal{R},n}^M$	spaces of spherical polynomials, p. 126
$\mathbb{S}_n^M, \mathbb{S}_{m,n}^M, \mathbb{S}_{\mathcal{R},n}^M$	unit spheres in the spaces of spherical polynomials, p. 126
$\mathcal{L}_n^M, \mathcal{L}_n^{M,m}, \mathcal{L}_n^{M,\mathcal{R}}$	subsets of the unit spheres $\mathbb{S}_n^M, \mathbb{S}_{m,n}^M, \mathbb{S}_{\mathcal{R},n}^M$, p. 126
$\mathcal{P}_n^M, \mathcal{P}_{m,n}^M, \mathcal{P}_{\mathcal{R},n}^M$	optimally space localized spherical polynomials, p. 127

Bibliography

[1] ASKEY, R., AND BINGHAM, N. H. Gaussian processes on compact symmetric spaces. *Z. Wahrscheinlichkeitstheor. Verw. Geb. 37* (1976), 127–143.

[2] BERGER, M. *A Panoramic View of Riemannian Geometry.* Springer-Verlag, Berlin, 2003.

[3] BERGER, M., GAUDUCHON, P., AND MAZET, E. *Le Spectre d'une Variété Riemannienne.* Lecture Notes in Mathematics 194, Springer-Verlag, Berlin-Heidelberg-New York, 1974.

[4] BESSE, A. L. *Manifolds all of whose Geodesics are Closed.* Springer-Verlag, Berlin-Heidelberg-New York, 1978.

[5] BISHOP, R. L., AND CRITTENDEN, R. J. *Geometry of Manifolds.* Academic Press, New York, 1964.

[6] BREITENBERGER, E. Uncertainty measures and uncertainty relations for angle observables. *Found. Phys. 15* (1983), 353–364.

[7] BUCHNER, M. A. Simplicial structure of the real analytic cut locus. *Proc. Am. Math. Soc. 64*, 1 (1977), 118–121.

[8] CHAVEL, I. *Eigenvalues in Riemannian Geometry.* Academic Press, Orlando, 1984.

[9] CHAVEL, I. *Riemannian Geometry: a Modern Introduction*, second ed. Cambridge University Press, Cambridge, 2006.

[10] CHIHARA, T. S. *An Introduction to Orthogonal Polynomials.* Gordon and Breach, Science Publishers, New York, 1978.

[11] CIATTI, P., RICCI, F., AND SUNDARI, M. Heisenberg-Pauli-Weyl uncertainty inequalities and polynomial growth. *Adv. Math. 215*, 2 (2007), 616–625.

[12] DO CARMO, M. P. A. *Riemannian Geometry.* Birkhäuser, Boston, 1992.

[13] DRIVER, K., JORDAAN, K., AND MBUYI, N. Interlacing of the zeros of Jacobi polynomials with different parameters. *Numer. Algor. 49*, 1-4 (2008), 143–152.

[14] DUNKL, C. F., AND XU, Y. *Orthogonal Polynomials of Several Variables.* Cambridge University Press, Cambridge, 2001.

[15] ERB, W. Uncertainty principles on compact Riemannian manifolds. *Appl. Comput. Harmon. Anal.* (2009). doi:10.1016/j.acha.2009.08.012.

[16] ERB, W., AND FILBIR, F. Approximation by positive definite functions on compact groups. *Numer. Funct. Anal. Optim. 29,* 9-10 (2008), 1082–1107.

[17] FILBIR, F., MHASKAR, H. N., AND PRESTIN, J. On a filter for exponentially localized kernels based on Jacobi polynomials. *J. Approximation Theory* (2009). doi: 10.1016/j.jat.2009.01.004.

[18] FISCHER, G. *Qualitative Unschärferelationen und Jacobi-Polynome.* Dissertation, Technische Universität München, 2005.

[19] FOLLAND, G. *Harmonic Analysis in Phase Space.* Princeton University Press, 1989.

[20] FOLLAND, G. B., AND SITARAM, A. The uncertainty principle: a mathematical survey. *J. Fourier Anal. Appl. 3,* 3 (1997), 207–233.

[21] FREEDEN, W., GERVENS, T., AND SCHREINER, M. *Constructive Approximation on the Sphere with Applications to Geomathematics.* Oxford Science Publications, Clarendon Press, Oxford, 1998.

[22] FREEDEN, W., AND WINDHEUSER, U. Combined spherical harmonic and wavelet expansion - a future concept in earth's gravitational determination. *Appl. Comput. Harmon. Anal 4,* 1 (1997), 1–37.

[23] GALLOT, S., HULIN, D., AND LAFONTAINE, J. *Riemannian Geometry.* Springer-Verlag, Berlin-Heidelberg, 1987.

[24] GANGOLLI, R. Positive-definite kernels on homogeneous spaces and certain stochastic processes related to Levy's Brownian motion of several parameters. *Ann. Inst. Henri Poincaré, Nouv. Sér., Sec. B 3,* 2 (1967), 121–225.

[25] GAUTSCHI, W. *Orthogonal Polynomials: Computation and Approximation.* Oxford University Press, Oxford, 2004.

[26] GOH, S. S., AND GOODMAN, T. N. Uncertainty principles and optimality on circles and spheres. In *Advances in constructive approximation: Vanderbilt 2003. Proceedings of the international conference, Nashville, TN, USA, May 14–17, 2003* (2004), M. Neamtu and E. B. Saff, Eds., Nashboro Press, Brentwood, TN, Modern Methods in Mathematics, pp. 207–218.

[27] GOH, S. S., AND GOODMAN, T. N. T. Uncertainty principles and asymptotic behavior. *Appl. Comput. Harmon. Anal. 16*, 1 (2004), 69–89.

[28] GOH, S. S., AND MICCHELLI, C. A. Uncertainty principles in Hilbert spaces. *J. Fourier Anal. Appl. 8*, 4 (2002), 335–373.

[29] GRÖCHENIG, K. *Foundations of Time-Frequency Analysis.* Birkhäuser, Boston, 2001.

[30] GROMOLL, D., KLINGENBERG, W., AND MEYER, W. *Riemannsche Geometrie im Großen.* Lecture Notes in Mathematics 55, Springer-Verlag, Berlin-Heidelberg-New York, 1968.

[31] HARDY, G. *Divergent Series.* Oxford University Press, Oxford, 1949.

[32] HAVIN, V., AND JÖRICKE, B. *The Uncertainty Principle in Harmonic Analysis.* Springer-Verlag, Berlin, 1994.

[33] HEISENBERG, W. Über den anschaulichen Inhalt der quantentheoretischen Kinematik und Mechanik. *Z. f. Physik 43* (1927), 172–198.

[34] HELGASON, S. *Differential Geometry and Symmetric Spaces.* Academic Press, New York-London, 1962.

[35] HELGASON, S. The Radon transform on Euclidean spaces, compact two-point homogeneous spaces and Grassmann manifolds. *Acta Math. 113* (1965), 153–180.

[36] HELGASON, S. *Differential Geometry, Lie Groups and Symmetric Spaces.* Academic Press, Orlando, Florida, 1978.

[37] HELGASON, S. *Groups and Geometric Analysis.* Academic Press, Orlando, Florida, 1984.

[38] HEWITT, E., AND STROMBERG, K. *Real and Abstract Analysis.* Springer-Verlag, Berlin-Heidelberg-New York, 1965.

[39] HORN, R. A., AND JOHNSON, C. R. *Matrix Analysis.* Cambridge University Press, Cambridge, 1985.

[40] IFANTIS, E. K., KOKOLOGIANNAKI, C. G., AND SIAFARIKAS, P. D. Newton sum rules and monotonicity properties of the zeros of scaled co-recursive associated polynomials. *Methods Appl. Anal. 3*, 4 (1996), 486–497.

[41] ISMAIL, M. E. The variation of zeros of certain orthogonal polynomials. *Adv. Appl. Math. 8* (1987), 111–118.

[42] ISMAIL, M. E. *Classical and Quantum Orthogonal Polynomials in One Variable*. Cambridge University Press, Cambridge, 2005.

[43] ITOH, J.-I., AND TANAKA, M. The Lipschitz continuity of the distance function to the cut locus. *Trans. Am. Math. Soc. 353*, 1 (2001), 21–40.

[44] JORDAAN, K., AND TOÓKOS, F. Interlacing theorems for the zeros of some orthogonal polynomials from different sequences. *Appl. Numer. Math. 59*, 8 (2009), 2015–2022.

[45] KENNARD, E. H. Zur Quantenmechanik einfacher Bewegungstypen. *Z. f. Physik 44* (1927), 326–352.

[46] KOORNWINDER, T. The addition formula for Jacobi polynomials and spherical harmonics. *SIAM J. appl. Math. 25* (1973), 236–246.

[47] LAÍN FERNÁNDEZ, N. *Polynomial Bases on the Sphere*. Dissertation, University of Lübeck, 2003.

[48] LAÍN FERNÁNDEZ, N. Optimally space-localized band-limited wavelets on $\mathbb{S}^{q-1}$. *J. Comput. Appl. Math. 199*, 1 (2007), 68–79.

[49] LANG, S. *Differential and Riemannian Manifolds*. Springer-Verlag, New York, 1995.

[50] LASSER, R. *Introduction to Fourier Series*. Marcel Dekker, New York, 1996.

[51] LETESSIER, J. Some results on co-recursive associated Laguerre and Jacobi polynomials. *SIAM J. Math. Anal. 25*, 2 (1994), 528–548.

[52] LETESSIER, J. On co-recursive associated Jacobi polynomials. *J. Comput. Appl. Math. 57*, 2 (1995), 203–213.

[53] LEVESLEY, J., AND RAGOZIN, D. L. The density of translates of zonal kernels on compact homogeneous spaces. *J. Approximation Theory 103* (2000), 252–268.

[54] LI, Z., AND LIU, L. Uncertainty principles for Jacobi expansions. *J. Math. Anal. Appl. 286*, 2 (2003), 652–663.

[55] LI, Z., AND LIU, L. Uncertainty principles for Sturm-Liouville operators. *Constructive Approximation 21*, 2 (2005), 193–205.

[56] MARTINI, A. Generalized uncertainty inequalities. *Math. Z.* (2009). doi:10.1007/s00209-009-0544-5.

[57] MEANEY, C. Localization of spherical harmonic expansions. *Monatsh. Math. 98* (1984), 65–74.

[58] MHASKAR, H. N., AND PRESTIN, J. Polynomial frames: a fast tour. In *Approximation Theory XI. Gatlinburg, 2004* (2005), C. K. Chui, L. L. Schumaker, and M. Neamtu, Eds., Nashboro Press, Brentwood, TN.

[59] MHASKAR, H. N., AND PRESTIN, J. Polynomial operators for spectral approximation of piecewise analytic functions. *Appl. Comput. Harmon. Anal. 26* (2009).

[60] MÜLLER, C. *Spherical Harmonics*. Lecture Notes in Mathematics 17, Springer-Verlag, Berlin-Heidelberg-New York, 1966.

[61] MULDOON, M. E. Properties of zeros of orthogonal polynomials and related functions. *J. Comput. Appl. Math. 48*, 1-2 (1993), 167–186.

[62] NARCOWICH, F. J., AND WARD, J. D. Nonstationary wavelets on the m-sphere for scattered data. *Appl. Comput. Harmon. Anal. 3*, 4 (1996), 324–336.

[63] NARCOWICH, F. J., AND WARD, J. D. Wavelets associated with periodic basis functions. *Appl. Comput. Harmon. Anal. 3*, 1 (1996), 40–56.

[64] PAPOULIS, A. *Probability, Random Variables, and Stochastic Processes*, third ed. McGraw-Hill, New York, 1991.

[65] PATI, V., SITARAM, A., SUNDARI, M., AND THANGAVELU, S. An uncertainty principle for eigenfunction expansions. *J. Fourier Anal. Appl. 2*, 5 (1996), 427–433.

[66] PETRUSHEV, P., AND XU, Y. Localized polynomial frames on the interval with Jacobi weights. *J. Fourier Anal. Appl. 11*, 5 (2005), 557–575.

[67] PRESTIN, J., AND QUAK, E. Time frequency localization of trigonometric Hermite operators. In *Approximation theory VIII. Vol. 2. Wavelets and multilevel approximation. Papers from the 8th Texas international conference, College Station, TX, USA, January 8-12, 1995.* (1995), C. K. Chui and L. L. Schumaker, Eds., World Scientific. Ser. Approx. Decompos. 6, Singapore, pp. 343–350.

[68] PRESTIN, J., AND QUAK, E. Optimal functions for a periodic uncertainty principle and multiresolution analysis. *Proc. Edinburgh Math. Soc. 42*, 2 (1999), 225–242.

[69] PRESTIN, J., QUAK, E., RAUHUT, H., AND SELIG, K. On the connection of uncertainty principles for functions on the circle and on the real line. *J. Fourier Anal. Appl. 9*, 4 (2003), 387–409.

[70] RAUHUT, H. An uncertainty principle for periodic functions. Diploma thesis, Technische Universität München, 2001. Available at *http://rauhut.ins.uni-bonn.de*.

[71] RAUHUT, H. Best time localized trigonometric polynomials and wavelets. *Adv. Comput. Math. 22*, 1 (2005), 1–20.

[72] RICCI, F. Uncertainty inequalities on spaces with polynomial volume growth. *Rend. Accad. Naz. Sci. XL Mem. Mat. Appl. (5) 29*, 1 (2005), 327–337.

[73] RÖSLER, M., AND VOIT, M. An uncertainty principle for ultraspherical expansions. *J. Math. Anal. Appl. 209* (1997), 624–634.

[74] RÖSLER, M., AND VOIT, M. An uncertainty principle for Hankel transforms. *Proc. Amer. Math. Soc. 127* (1999), 183–194.

[75] RUDIN, W. *Real and Complex Analysis.* MacGraw-Hill, New York, 1970.

[76] RUDIN, W. *Functional Analysis.* MacGraw-Hill, New York, 1973.

[77] SCHWARTZ, L. *Cours d'Analyse de l'Ecole Polytechnique.* Hermann, Paris, 1967.

[78] SELIG, K. K. Trigonometric wavelets and the uncertainty principle. In *Approximation Theory, Proceedings of the 1st international Dortmund meeting IDoMAT 95 held in Witten, Germany, March 13-17, 1995*, M.W. Müller and M. Felten and D.H. Mache, Ed., Akademie Verlag. Math. Res. 86, Berlin.

[79] SELIG, K. K. Uncertainty principles revisited. *ETNA, Electron. Trans. Numer. Anal. 14* (2002), 164–176.

[80] SHERMAN, T. O. The Helgason Fourier transform for compact Riemannian symmetric spaces of rank one. *Acta Math. 164*, 1-2 (1990), 73–144.

[81] SIAFARIKAS, P. D. Inequalities for the zeros of the associated ultraspherical polynomials. *Math. Inequal. Appl. 2*, 2 (1999), 233–241.

[82] SUN, L. An uncertainty principle on hyperbolic space. *Proc. Am. Math. Soc. 121*, 2 (1994), 471–479.

[83] SZEGÖ, G. *Orthogonal Polynomials.* American Mathematical Society, Providence, Rhode Island, 1939.

[84] WANG, H.-C. Two-point homogeneous spaces. *Ann. Math. 55*, 2 (1952), 177–191.

[85] WERNER, D. *Funktionalanalysis*, forth ed. Springer-Verlag, Berlin, 2002.

[86] WEYL, H. *The Theory of Groups and Quantum Mechanics.* Dover Publications, New York, translated from the second german edition from H.P. Robertson, 1931.

[87] WIMP, J. Explicit formulas for the associated Jacobi polynomials and some applications. *Can. J. Math. 39* (1987), 983–1000.

[88] ZIEMER, W. P. *Weakly Differentiable Functions - Sobolev Spaces and Spaces of Bounded Variation.* Springer-Verlag, New York, 1989.